INTRODUCTION TO

SEMICONDUCTOR PHYSICS

INTRODUCTION TO SEMICONDUCTOR PHYSICS

Holger T. Grahn

Paul-Drude-Institut für Festkörperelektronik
Berlin, Germany

Published by

World Scientific Publishing Co. Pte. Ltd.
P O Box 128, Farrer Road, Singapore 912805
USA office: Suite 1B, 1060 Main Street, River Edge, NJ 07661
UK office: 57 Shelton Street, Covent Garden, London WC2H 9HE

British Library Cataloguing-in-Publication Data
A catalogue record for this book is available from the British Library.

First published 1999
Reprinted 2001

INTRODUCTION TO SEMICONDUCTOR PHYSICS

ISBN 981-02-3302-7

This book is printed on acid-free paper.

Printed in Singapore by Uto-Print

PREFACE

Semiconductors have become increasingly important over the last fifty years since the invention of the transistor, which was originally based on the semiconducting material germanium. However, today silicon is by far the technologically most important semiconductor. The development of the information highway is closely connected with the success of the miniaturization of silicon-based electronics. By the year 2010, dynamical random access memory (DRAM) chips with 10 Gigabyte memory are expected to be incorporated in conventional personal computers. In the last 15 years, additional applications of semiconductors such as light detectors (solar cells) and light emitters (light emitting diodes and lasers) have become increasingly important. In contrast to electronic devices, light-emitting devices are based on direct energy gap semiconductors such as GaAs and related compounds, ZnSe, and most recently GaN.

The first part of this book is based on an international course entitled *Introduction to Semiconductor Physics* given within the Department of Physical Electronics at Tōkyō Institute of Technology (Tōkyō Kōgyō Daigaku) in Tōkyō, Japan, during the second semester of the academic year 1995-96. The second part, in particular the extensive discussion of optical properties including the effects of external fields, is based on a lecture entitled *Optical Properties of Semiconductors* given within the Institute for Solid State Physics (Institut für Festkörperphysik) at the Technical University (Technische Universität) Berlin, Germany, during the spring semester 1997. This introductory course does not require a deep knowledge of condensed matter physics. However, a basic knowledge of classical physics, quantum mechanics, and condensed matter physics is essential in order to follow the material presented here. This book also contains a summary of many material parameters of the most commonly used and investigated semiconductors. Since band structure parameters such as effective masses as well as optical constants, e. g., the dielectric constant, are sometimes revised in the course of time, not all parameters listed in this book correspond to their most recent values. For this introductory course, the parameters are only used to show general trends and to provide a sense for their possible range.

After an introduction, the book begins with a review of the crystal structure of semiconductors followed by a chapter on the formation of energy bands and energy gaps, which is discussed within the framework of the periodic potential. The band structure of the most common semiconductors together with the concept of the effective mass is presented in the following chapter. After introducing the density of states and their critical points for bulk as well as for low-dimensional semiconductors, the statistics of carriers and the temperature dependence of the carrier density are reviewed. The next chapter deals with basic models for carrier transport and the Hall effect. Before discussing the important scattering processes, which determine the carrier mobility, the phonon dispersion of semiconductors is presented followed by the phonon statistics and the thermal occupation of phonon states. After reviewing the scattering processes, the concept of excitons is introduced, which forms the basis for the second part of the book focusing on optical properties. In the last three chapters, the optical absorption of free carriers and excitons as well as emission processes are presented for bulk and low-dimensional semiconductors including the effects of an external electric and magnetic field.

I would like to thank in particular the students of both courses, who encouraged me to supply them with notes in form of a manuscript. These scripts lay the foundation for this book. In the final stages of the preparation of the manuscript, I received great support from O. Brandt, K.J. Luo, and P. Santos at the Paul-Drude-Institute as well as E. Runge from Humboldt University in Berlin, who all made important suggestions for the improvement of the manuscript. I also would like to acknowledge the support and hospitality received during my sabbatical in Japan, in particular from Prof. Kiyoshi Takahashi (Teikyō University of Science and Technology) and Prof. Makoto Konagai (Tōkyō Institute of Technology). Finally, my wife deserves great admiration for her patience during the final completion of the manuscript.

Berlin, October 1998 Holger Grahn

Acknowledgements for revised edition:

I am very grateful to all the people, who have informed me about errors and misprints in the first printing of this book, in particular S. Teitsworth (Duke University) and D. Jena (University of California at Santa Barbara). During the corrections of the manuscript, I received great support from H.-Y. Hao and M. Rogozia at the Paul-Drude-Institute. Finally, I would like to acknowledge the support and hospitality of Prof. Kiyoshi Takahashi (Teikyō University of Science and Technology) during my recent visit to Japan.

Berlin, January 2001 Holger Grahn

CONTENTS

CHAPTER 1
INTRODUCTION

Over the last 50 years, semiconductors have become the most important material for the fabrication of electronic and optoelectronic devices. In particular, the invention of the transistor 50 years ago in December 1947 by J. Bardeen, W.H. Brattain, and W.B. Shockley has resulted in the development of the so-called information society of today. Although the first transistor was made from germanium (Ge), today only one semiconductor, namely silicon (Si), dominates the production of transistors and integrated circuits. Both, Si and Ge, are elementary semiconductors from the fourth group of the periodic table. However, there is a much larger number of semiconductors available today, and more and more III-V and II-VI compound semiconductors find their application. Within the last decade we have seen GaAs high-electron mobility transistors used in satellite dishes as well as cellular phones and $GaAs/Al_xGa_{1-x}As$ laser diodes in compact disc players. A blue laser based on ZnSe was realized some years ago. Very recently a blue light-emitting laser diode based on GaN was developed. Finally, solar cells are fabricated from semiconducting materials. Their use in houses and consumer products is steadily increasing.

The trend of the computer industry to achieve larger and larger scale integration, such as the 256-megabit chip DRAM, which is going into production in 1998, and the discovery of the quantum Hall effect in 1980 have resulted in the quest for new electronic and optoelectronic devices based on low-dimensional semiconductor structures such as two-, one- and even zero-dimensional systems. In these structures, quantum mechanical effects will become essential, once a typical length scale of 100 nm or less is reached. Currently, researchers around the world are aiming to realize quantum wire and quantum dot lasers as well as devices based on single electron tunneling. The future will tell us, which of these devices will become essential for consumer products.

1. What is a Semiconductor?

There are several ways of defining a semiconductor. Historically, the term semiconductor has been used to denote materials with a much higher conductivity than insulators, but a much lower conductivity than metals measured at room temperature. Today there are two more types of conductors: superconductors and semimetals. Typical conductivities of superconductors, metals, semimetals, semiconductors and insulators are listed in Tab. 1.1.

This definition is not complete. What really distinguishes metals from semiconductors is the temperature dependence of the conductivity. While metals (except for superconductors) and semimetals retain their metallic conductivity even at low temperatures, semiconductors are transformed into insulators at very low temperatures. In this sense semiconductors and insulators are actually one class of materials, which differs from metals and semimetals, which form another class. This classification is directly connected to the existence of a gap between occupied and empty states, i.e., an energy gap, in semiconductors and insulators. In Tab. 1.2 the classification according to the energy gap E_G is summarized.

The border line between semiconductors and insulators is rather arbi-

Table 1.1. Typical conductivities ($\sigma = \frac{l}{RA}$, where l denotes the length, R the resistance, and A the cross sectional area of the conductor) of superconductors, metals, semimetals, semiconductors, and insulators at room temperature.

Type of solid	σ (Ω^{-1} cm^{-1})	Example
Superconductor (low temperature)	$> 10^{10}$	Pb, $YBa_2Cu_3O_7$
Metal	$10^5 - 10^{10}$	Au, Cu, Pb, Ag
Semimetal	$10^2 - 10^5$	graphite (C), HgTe
Semiconductor	$10^{-9} - 10^2$	Si, Ge, GaAs, InSb, ZnSe
Insulator	$< 10^{-9}$	quartz (SiO_2), CaF_2

Table 1.2. Classification of solids according to their energy gap E_G and carrier density n at room temperature.

Type of solid	E_G (eV)	n (cm^{-3})
Metal	no energy gap	10^{22}
Semimetal	$E_G \leq 0$	$10^{17} - 10^{21}$
Semiconductor	$0 < E_G < 4$	$< 10^{17}$
Insulator	$E_G \geq 4$	$\ll 1$

trary. In particular, the value of the energy gap separating the semiconducting materials from the insulating ones is not well-defined. For example, diamond (C) was considered for a long time an insulator, but today it is possible to prepare it in such a way that it has semiconducting properties even at room temperature. The important distinction between these two systems originates historically from their different conductivities at room temperature. However, an insulator at room temperature can become a semiconductor at higher temperatures. Therefore, wide energy gap materials are currently under investigation for high temperature electronics.

Another possibility of defining a semiconductor, which is related to the energy gap, is through the free carrier concentration at room temperature. Metals and semimetals have a rather large carrier density, semiconductors exhibit a moderate carrier density at room temperature, while insulators have a negligible carrier density. Typical carrier densities for these different types of solids are compiled in Tab. 1.2. The listed densities are intrinsic values, i.e., for pure materials. However, real semiconductors always contain some impurities, which can act as dopants leading to larger values for the carrier densities than the intrinsic ones.

To summarize, a semiconductor is a solid with a finite energy gap below 4 eV, which results in a moderate conductivity and carrier density at room temperature. By doping the semiconductor in a controlled fashion, the conductivity and carrier density can be varied over several orders of magnitude. This is not possible in metals, which always exhibit a large conductivity and large carrier density. Due to the existence of the energy gap, semiconductors are transparent for energies below the gap, i.e., in the far- to near-infrared region depending on the value of the energy gap. However, they strongly absorb light for energies above the energy gap, typically in the near-infrared to visible regime. This behavior again is in strong contrast to metals, which are usually opaque from the far-infrared to the visible regime. In the absorptive region, the conductivity of semiconductors increases, when they are irradiated.

2. Classification of Semiconductors

There is a large variety of semiconductors available today, although for applications Si completely dominates the market. Nevertheless, other semiconductors can have quite different properties. For example, Si cannot be used for light emitting diodes or lasers, while semiconductors such as GaAs can. In this section, semiconductors will be classified according to their chemical composition. There are elementary semiconductors such as Si, Ge, and gray tin (α-Sn), which all belong to group IV in the periodic table. Therefore, these systems are usually referred to as group IV semiconductors. Another group IV element is carbon, which solidifies into two structures, diamond and graphite. Diamond is an insulator and has the same crystal structure as Si, Ge, and α-Sn. Graphite is a semimetal and exhibits a hexagonal structure. Sn also exists in two phases, white tin (β-Sn), which is metallic, and semiconducting α-Sn. The last element of the group IV is lead, which is metallic. All group IV elements are listed in Tab. 1.3 including their lattice constant, energy gap, and lattice structure.

Group IV elements are exceptional in the periodic table in the sense that the outer shell of the individual atoms is exactly half filled. By sharing one of the four electrons of the outer shell with another Si atom, a three-dimensional crystal structure with no preferential direction (except for graphite) can be realized. One can also combine two different group IV semiconductors to obtain a compound material such as SiC or SiGe. SiC is a material close to the border line between semiconductors and insulators with a lattice constant of 0.436 nm and an energy gap of 3.0 eV (413 nm). This material is investigated for high-temperature electronic applications.

The elementary group IV semiconductors can also be understood as a special case of the 8N rule, i.e., completing the outer shell by sharing electrons with other atoms. There are many other ways to fulfill this 8N rule. However, these materials are all compound semiconductors. Ele-

Table 1.3. Lattice constant a, energy gap E_G at 300 K, type of energy gap, and lattice structure of group IV elements.

Material	a (nm)	E_G (eV)	E_G (nm)	Type	Structure
Diamond (C)	0.357	5.48	226	indirect	cubic
Silicon (Si)	0.543	1.12	1107	indirect	cubic
Germanium (Ge)	0.566	0.664	1867	indirect	cubic
Gray tin (α-Sn)	0.649	—	—	—	cubic
White tin (β-Sn)	0.583 0.318	—	—	—	tetragonal
Graphite (C)	0.246 0.673	—	—	—	hexagonal
Lead (Pb)	0.495	—	—	—	cubic

ments from group III (II) can be combined with group V (VI) elements. Group I elements in conjunction with group VII elements lead to wide energy gap insulators, since these materials are formed by ionic bonds and not covalent bonds as III-V and most II-VI semiconductors. Most of the III-V semiconductors exist in the so-called zincblende structure, which is a cubic lattice. Some exist in the wurtzite structure, which corresponds to a hexagonal lattice. GaAs is the best-known III-V compound semiconductor, while GaN, although known for a long time, has only recently become very important stimulating a lot of research on this material. In contrast to Si, GaAs, and many other compound semiconductors are direct semiconductors so that optical applications of these systems are very common. The most common III-V semiconductors are compiled in Tab. 1.4 together with their respective lattice constant, energy gap, and lattice structure.

One can combine for example GaAs and AlAs to obtain a ternary III-V compound $Al_xGa_{1-x}As$, which is a mixed crystal. One can also com-

Table 1.4. Lattice constant a, energy gap E_G at 300 K, type of energy gap, and lattice structure of III-V semiconductors.

Material	a (nm)	E_G (eV)	E_G (nm)	Type	Structure
BN	0.362	6.4	194	indirect	cubic
	0.666, 0.250	5.2	238	direct	hexagonal
AlN	0.311, 0.498	6.2	200	direct	hexagonal
GaN	0.318, 0.517	3.44	360	direct	hexagonal
AlP	0.546	2.51	494	indirect	cubic
BP	0.454	2.4	517	indirect	cubic
GaP	0.545	2.27	546	indirect	cubic
AlAs	0.566	2.15	577	indirect	cubic
InN	0.354, 0.870	1.89	656	direct	hexagonal
AlSb	0.614	1.62	765	indirect	cubic
GaAs	0.565	1.424	871	direct	cubic
InP	0.587	1.34	925	direct	cubic
GaSb	0.610	0.75	1653	direct	cubic
InAs	0.606	0.354	3502	direct	cubic
InSb	0.648	0.18	6890	direct	cubic

bine four elements, two from group III and two from group V, to form quaternary compounds such as $In_xGa_{1-x}As_{1-y}P_y$. There is a large variety of materials available today using III-V compound semiconductors. III-V semiconductors are characterized by covalent bonding. However, in contrast to group IV semiconductors, III-V compounds exhibit a certain degree of ionic bonding. In GaAs for example, the bonding has an ionic character of about 30%, while NaCl (a typical I-VII compound) for comparison has an ionic character of 94%, since it is dominated by ionic bonding.

Another class of semiconductors are II-VI semiconductors. The most well-known materials are listed in Tab. 1.5 together with their lattice constant, energy gap and lattice structure. II-VI compounds typically exhibit a larger degree of ionic bonding than III-V compounds, since the respec-

Table 1.5. Lattice constant a, energy gap E_G at 300 K (* denotes value at low temperature), type of energy gap, and lattice structure of II-VI semiconductors.

Material	a (nm)	E_G (eV)	E_G (nm)	Type	Structure
ZnS	0.541	3.68	337	direct	cubic
	0.382, 0.626	3.91	317	direct	hexagonal
ZnO	0.325, 0.521	3.44*	360	direct	hexagonal
ZnSe	0.567	2.7	459	direct	cubic
	0.40, 0.654			direct	hexagonal
CdS	0.582	2.55	486	direct	cubic
	0.414, 0.671	2.51	494	direct	hexagonal
ZnTe	0.610	2.28	544	direct	cubic
α-HgS	0.415, 0.950	2.1	590	direct	trigonal
β-HgS	0.585	—	—	—	cubic
CdSe	0.605	1.9*	653	direct	cubic
	0.43, 0.701	1.75	709	direct	hexagonal
CdTe	0.648	1.475	841	direct	cubic
HgSe	0.609	—	—	—	cubic
HgTe	0.646	—	—	—	cubic

tive elements differ more in the electron affinity due to their location in the periodic table. In ZnO, where the covalent and ionic character are almost balanced, the ionic character is about 60%.

I-VII compounds can also form semiconductors, which can exhibit a very large ionicity. The energy gap is considerably larger than that in many III-V compounds. The most important materials of this type are listed in Tab. 1.6. The compounds AgCl and AgBr crystallize in the NaCl structure, which is also cubic, but differs from the zincblende structure.

In Fig. 1.1 the energy gaps of the listed group IV, III-V, II-VI, and I-VII semiconductors are shown versus the lattice constant for materials with a band gap below 4 eV (310 nm) and a lattice constant above 0.51 nm.

Table 1.6. Lattice constant a, energy gap E_G at low temperature, type of energy gap, and lattice structure of I-VII semiconductors.

Material	a (nm)	E_G (eV)	E_G (nm)	Type	Structure
γ-CuCl	0.541	3.395	365	direct	cubic
AgCl	0.555	3.249	382	indirect	cubic
γ-CuI	0.604	3.115	398	direct	cubic
γ-CuBr	0.569	3.077	403	direct	cubic
β-AgI	0.458, 0.749	3.024	410	direct	hexagonal
AgBr	0.577	2.684	462	indirect	cubic

There is a clear tendency for an increasing energy gap with increasing ionic bonding, i.e., going from group IV to III-V, to II-VI, and to I-VII semiconductors. There are other elementary semiconductors such as selenium and

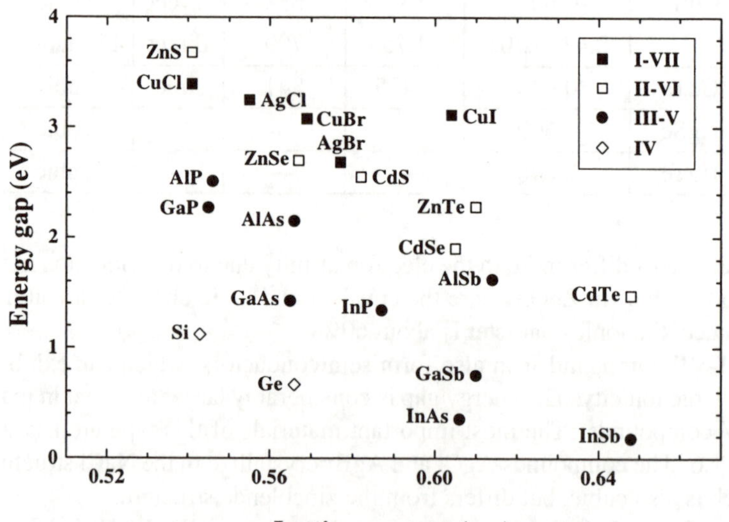

Fig. 1.1. Energy gap vs lattice constant in group IV, III-V, II-VI, and I-VII semiconductors.

Table 1.7. Lattice constant a, energy gap E_G (* denotes value at low temperature), and lattice structure of other semiconductors.

Material	Type	a (nm)	E_G (eV)	E_G (nm)	Structure
Se	VI	0.435,0.495	1.85*	670	trigonal
Te	VI	0.445	0.34	3760	chain
PbS	IV-VI	0.594	0.41	3020	cubic
PbTe	IV-VI	0.646	0.31	4000	cubic
PbSe	IV-VI	0.612	0.28	4430	cubic
Bi_2Te_3	V-VI	0.438	0.13	9540	trigonal
CdSb	II-V	0.647	0.49	2580	orthorhombic
Cd_3As_2	II-V	1.26, 2.54	–0.19	—	tetragonal

tellurium from group VI, the chalcogenes. However, since group VI elements have only two missing valence electrons to be shared with neighboring atoms, these materials have a strong tendency to form chain structures. Other compound semiconductors include the IV-VI compounds PbS, PbSe, and PbTe. A combination of group V and VI is also possible, e.g., Bi_2Te_3. Another class are II-V compounds such as Cd_3As_2 and CdSb. In Tab. 1.7 the lattice constant, energy gap, and structure of these other semiconductors are compiled. Even some of the high-temperature superconductors are actually in the normal state semiconducting and not metallic.

Finally, there is a number of amorphous semiconductors. The most well-known is a mixture of Si and hydrogen (H), usually referred to as amorphous hydrogenate silicon (a-Si:H). Furthermore, there are chalcogenide glasses such as As_2Te_3 and As_2Se_3. Some chalcogene elements as well as chalcogenides glasses such as As_2Se_3 are used in xerography.

3. A Brief History of Semiconductor Physics

In 1879, E.H. Hall discovered an effect, which was essential for the development of methods to characterize semiconductors. The Hall effect allows the determination of the carrier density as well as the type of the charge carriers. J.J. Thomson's discovery of the electron in 1897 stimulated the development of theoretical descriptions of the conduction in metals. Three years later, P. Drude developed a model, which describes the electrical and thermal conduction of solids. However, a quantitative description of carrier transport in semiconductors was only possible after quantum mechanics was available and applied to solids. In 1926, F. Bloch formulated a quantum mechanical theorem, now referred to as Bloch's theorem, which describes the electron wave function taking into account the crystal structure of the solid. Another important discovery was made by A.H. Wilson in 1931, who demonstrated that semiconductors are insulators with narrow energy gaps. He also introduced the concept of holes. The energy gap is also important for understanding the reverse breakdown of semiconductors and insulators in large electric fields. In 1934, C. Zener showed that it originates from interband tunneling, i.e., tunneling through the energy gap.

Fifty years ago, Bardeen, Brattain, and Shockley developed the first transistor opening the way for semiconductor devices. The first semiconductor lasers were fabricated in the early 1960's. A vast amount of research on III-V and II-VI semiconductors in the following decades was stimulated by the development of epitaxial growth techniques in the late 1960's to early 1970's, in particular for heterostructures, which are composed of two or more semiconductors with different energy gaps. A new wave of research on low-dimensional semiconductor heterostructures was triggered by the discovery of the quantum Hall effect in 1980 by K. von Klitzing in the inversion layer of a field effect transistor, which forms a quasi two-dimensional layer. This research is geared today towards the realization of quantum wire and dot lasers as well as a single electron transistor. A missing link for optoelectronic applications was discovered in 1992, the blue laser based on II-VI semiconductors. However, presently its lifetime is still rather short. In 1993, the blue light-emitting diode (LED) based on GaN was demonstrated, which is now already in mass production. In 1997, the first blue laser diode based on GaN with an estimated lifetime of more than 10,000 hours was reported. These are only a few highlights of semiconductor physics. I have to apologize, if I left out an important development.

CHAPTER 2

CRYSTAL STRUCTURE AND RECIPROCAL LATTICE

1. Principles of Crystal Lattices

In crystalline solids such as semiconductors, the individual atoms form a periodic array of sites called the crystal lattice. The word crystal originates from the Greek word *kristallos*, which stands for ice. For a solid such as Si, the equilibrium state is reached, when the environment viewed from each atom looks exactly the same. The resulting periodicity is very important in describing some basic properties of semiconductors. In amorphous semiconductors such as amorphous Si, the long range periodicity does not exist anymore. However, within a few lattice constants the structure looks almost as periodic as crystalline Si. Small deviations from the periodicity do not change the local arrangement, but only break the long-range periodicity.

The crystal lattice is typically described by a translation or Bravais lattice. The French natural scientist A. Bravais was the first to classify all possible 3-dimensional lattices according to their symmetry. A Bravais lattice consists of an array of points, which can be generated by the translation vector $\underline{R}$ given by

$$\underline{R} = \sum_{i=1}^{3} N_i \, \underline{a}_i \, . \tag{2.1}$$

Here, N_i denote integer numbers and $\underline{a}_i$ the three primitive vectors of the lattice, which are three vectors not all in the same plane. Each lattice point represents an atom or a group of atoms. The arrangement of these lattice points acquires a well-defined symmetry. The lattice point can be located at the center of an atom or of a group of atoms, but this is not required.

The volume defined by the three vectors $\underline{a}_i$ is called the primitive unit cell. By translating this unit cell through all the vectors in a Bravais lattice, all of space is filled without any overlap between this unit cells or any voids between them.

There is no unique way of choosing a primitive cell for a given Bravais lattice. An important primitive unit cell is the Wigner-Seitz cell, which exhibits the full symmetry of the Bravais lattice. It is defined by the region around the lattice point, which is closer to that point than to any other lattice point. It is usually constructed by (i) drawing lines connecting the lattice point to all others in the lattice, (ii) bisecting each line with a plane, and (iii) taking the smallest polyhedron bounded by these planes. An example of a Wigner-Seitz cell in two dimensions is shown by the shaded area in Fig. 2.1.

The three dimensional lattices with the highest degree of symmetry are the cubic lattices, which include the simple cubic (sc), the body-centered cubic (bcc), and the face-centered cubic lattice (fcc). Another important lattice with a smaller degree of symmetry is the hexagonal lattice. These four lattices are of particular importance for semiconductors, and we will therefore discuss these lattices briefly, before turning to semiconductor lattices.

The sc lattice can be described using the orthogonal vectors $\underline{a}_1 = a\hat{\underline{x}}$,

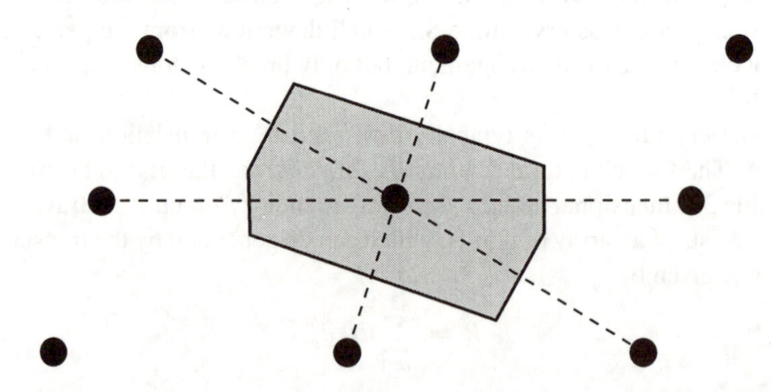

Fig. 2.1. The Wigner-Seitz cell (shaded area) for a two-dimensional Bravais lattice. In two dimensions the Wigner-Seitz cell is always a hexagon unless the lattice is rectangular.

$\underline{a}_2 = a\hat{\underline{y}}$, and $\underline{a}_3 = a\hat{\underline{z}}$, where $\hat{\underline{x}}$, $\hat{\underline{y}}$, and $\hat{\underline{z}}$ are three orthogonal unit vectors and a denotes the lattice constant. The number of nearest neighbors in the sc lattice is six. The Wigner-Seitz cell of the sc lattice is a cube. The other two cubic structures are the bcc and fcc lattices. In the bcc lattice, an additional lattice point is added at the center of the cube as shown in the top left panel of Fig. 2.2. A set of three primitive vectors consists of $\underline{a}_1 = a\hat{\underline{x}}$, $\underline{a}_2 = a\hat{\underline{y}}$, and $\underline{a}_3 = \frac{a}{2}(\hat{\underline{x}}+\hat{\underline{y}}+\hat{\underline{z}})$. For the bcc lattice the number of nearest neighbors is increased to eight. The Wigner-Seitz cell of the bcc lattice shown by the shaded area in the bottom left panel of Fig. 2.2 consists of a truncated octahedron, i.e., a regular octahedron (eight faces) with the

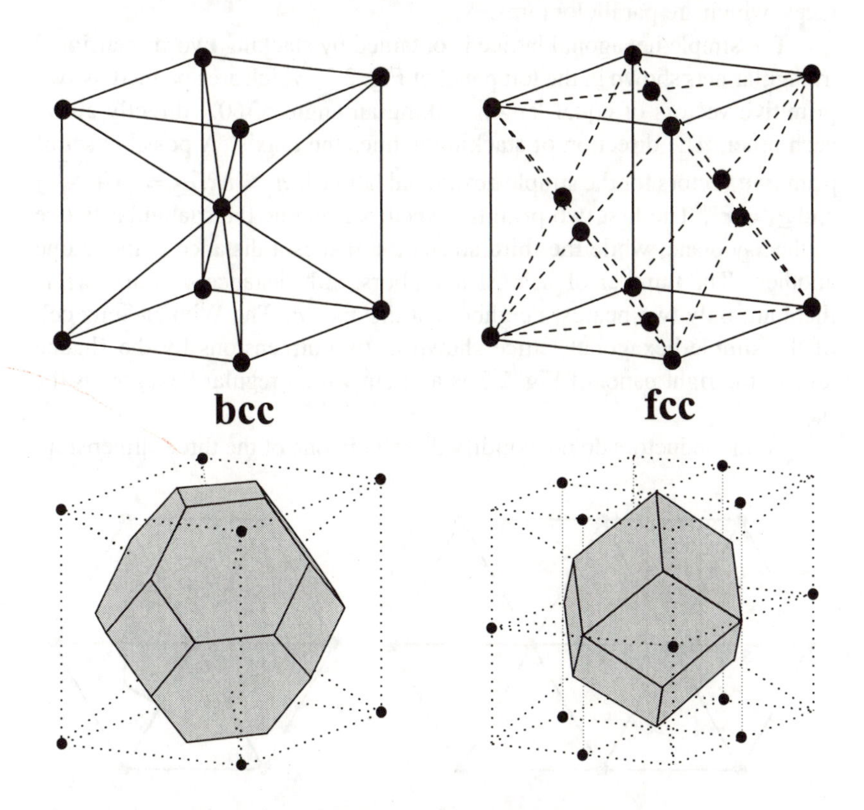

Fig. 2.2. The bcc and fcc lattices (top) and their Wigner-Seitz cells (bottom) indicated by the shaded areas.

tips in the directions of the three principal axes cut off forming squares (six faces). The fcc lattice can be obtained from the sc lattice by adding a lattice point at the center of each square face as shown in the top right panel of Fig. 2.2. A symmetric set of primitive vectors is $\underline{a}_1 = \frac{a}{2}(\hat{\underline{y}} + \hat{\underline{z}})$, $\underline{a}_2 = \frac{a}{2}(\hat{\underline{z}} + \hat{\underline{x}})$, and $\underline{a}_3 = \frac{a}{2}(\hat{\underline{x}} + \hat{\underline{y}})$. The number of nearest neighbors in the fcc lattice increases to twelve. Therefore, the packing density in the fcc lattice is about 1.41 times larger than that in the sc lattice and 1.09 times larger than that in the bcc lattice. The Wigner-Seitz cell of the fcc lattice shown by the shaded area in the bottom right panel of Fig. 2.2 consists of a rhombic dodecahedron, i.e., a geometric figure with twelve congruent faces, which are parallelograms.

The simple hexagonal lattice is obtained by stacking two-dimensional triangular nets shown in the left panel of Fig. 2.3, which are spanned by two primitive vectors of equal length making an angle of 60°, directly above each other. The direction of stacking defines the c-axis. A possible set of primitive vectors for the simple hexagonal lattice is $\underline{a}_1 = a\hat{\underline{x}}$, $\underline{a}_2 = \frac{a}{2}\hat{\underline{x}} + \frac{\sqrt{3}a}{2}\hat{\underline{y}}$ and $\underline{a}_3 = c\hat{\underline{z}}$. The first two primitive vectors generate the triangular lattice in the x-y plane, while the third stacks the planes a distance c above one another. The number of nearest neighbors with distance a is six, while there are only two nearest neighbors at distance c. The Wigner-Seitz cell of the simple hexagonal lattice shown in two dimensions by the shaded area in the right panel of Fig. 2.3 is a prism with a regular hexagon as the base.

Semiconductors do not solidify directly in one of the three-dimension-

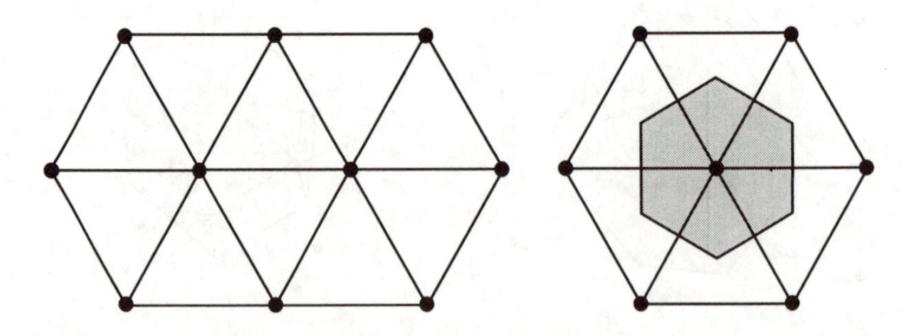

Fig. 2.3. The simple hexagonal lattice (left) and its Wigner-Seitz cell (right) indicated by the shaded area in two dimensions.

al cubic lattices such as the sc, bcc, or fcc lattice or in the simple hexagonal lattice. However, the fcc and the hexagonal lattices are the most important crystal lattices in connection with group IV, III-V, and II-VI semiconductors. In order to describe even elementary semiconductors, it is necessary to introduce the concept of a lattice with a basis. For example, the bcc lattice can be described as simple cubic lattices spanned by $a\underline{\hat{x}}$, $a\underline{\hat{y}}$, and $a\underline{\hat{z}}$ with a two point basis at

$$\underline{0},\ \frac{a}{2}\left(\underline{\hat{x}} + \underline{\hat{y}} + \underline{\hat{z}}\right). \tag{2.2}$$

For the fcc lattice, a four point basis at

$$\underline{0},\ \frac{a}{2}\left(\underline{\hat{x}} + \underline{\hat{y}}\right),\ \frac{a}{2}\left(\underline{\hat{y}} + \underline{\hat{z}}\right),\ \text{and}\ \frac{a}{2}\left(\underline{\hat{z}} + \underline{\hat{x}}\right) \tag{2.3}$$

is needed, when describing it by sc lattices. It is obvious that materials such as compound semiconductors have to be described through a lattice with a basis, since the unit cell has to contain at least two different atoms. However, even the elementary group IV semiconductors form a lattice with a basis, i.e., they cannot be described by a monatomic Bravais lattice. In contrast to semiconductors, many elementary metals solidify on monatomic Bravais lattices. The crystal structure of a solid therefore consists of identical copies of the same physical unit, called the basis, located at all the points of a Bravais lattice.

1.1. The diamond structure

Group IV elementary semiconductors such as Si, Ge, and α-Sn crystallize in the same structure as diamond. All these materials exhibit tetrahedral bonding form by sp^3-hybridization so that each atom is surrounded by four nearest neighbors located at the four corners of a regular tetrahedron. This configuration of neighbors cannot be represented by any of the basic cubic lattices. It can, however, be fulfilled by two fcc lattices with a basis as shown in Fig. 2.4, where one carbon atom resides at $\underline{0}$ and the second one at one-quarter of the body diagonal $\frac{a}{4}(\underline{\hat{x}}+\underline{\hat{y}}+\underline{\hat{z}})$. The lattice constant is 2.31 times the bond length. The number of nearest neighbors is four and the number of atoms per primitive cell two.

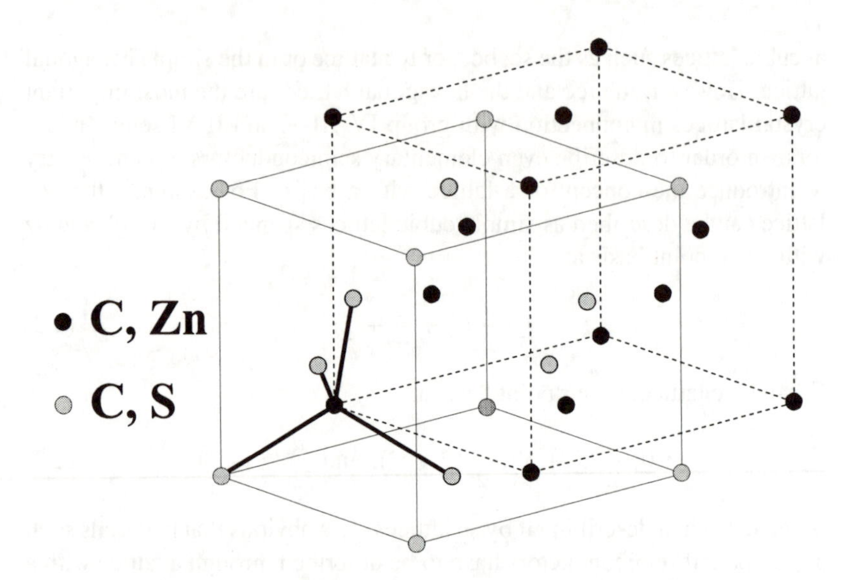

Fig. 2.4. The diamond crystal structure. If the two fcc lattices contain different atoms (black and gray circles), the zincblende structure is generated. One tetrahedral group is indicated by the thick lines.

1.2. The zincblende structure

The zincblende structure is named after the cubic phase of ZnS, in German Zinkblende (in English sphalerite). It corresponds to the diamond structure with two different atoms forming the basis, e.g., zinc and sulfur, as shown in Fig. 2.4. Each atom of one kind has four nearest neighbors of the opposite kind. This structure must be described by a lattice with a basis because of the geometrical position of the atoms and the two different types of atoms. One fcc sub-lattice contains the group-III (II) atoms, the other the group-V (VI) atoms. The cubic III-V and II-VI semiconductors in Tab. 1.5 and 1.6 crystallize in the zincblende structure. This structure is the most common lattice for binary compounds with covalent bonding. As in the diamond structure, the lattice constant is 2.31 times the bond length, the number of nearest neighbors is four, and the number of atoms per primitive cell is two.

1.3. The graphite and hexagonal close-packed structures

In order to understand the third important semiconductor crystal lattice, it is instructive to look at the other phase of carbon, namely graphite. In this phase, the bonding is dominated by sp^2-hybridization. Therefore, graphite exhibits a layered structure consisting of planes of hexagonal rings similar to those found in benzene. This two-dimensional structure, usually referred to as a honeycomb lattice, is not a Bravais lattice. In two consecutive layers, the hexagons are displaced as shown in Fig. 2.5(a) in such a way that only half the atoms occupy the same position in the plane, while the

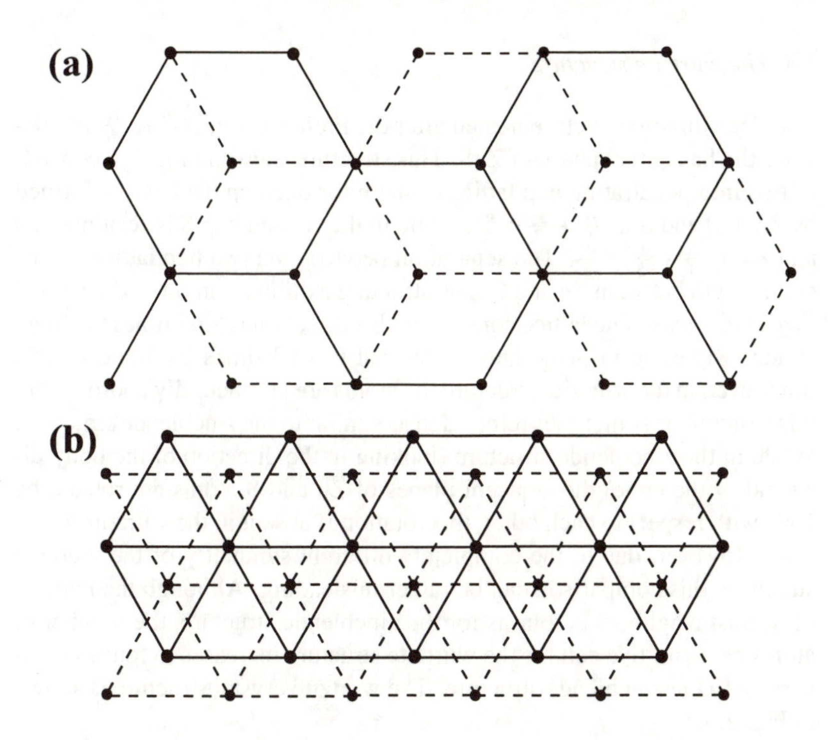

Fig. 2.5. The graphite (a) and the hexagonal close-packed (b) crystal structures. The solid and dashed lines denote planes separated by one-half of the lattice constant perpendicular to the hexagons (a) or hexagonal layers (b).

other half occupies positions corresponding to the center of the hexagons in the adjacent layers. The structure completely repeats itself after the second layer. The lattice constant perpendicular to the planes is about 4.8 times the the nearest-neighbor distance within the planes.

An even more important crystal lattice is the hexagonal close-packed (hcp) lattice, which is formed by two interpenetrating simple hexagonal Bravais lattices. The structure can be described by a hexagonal lattice with a basis consisting of one atom at $\underline{0}$ and the second one at $\frac{\underline{a}_1}{3} + \frac{\underline{a}_2}{3} + \frac{\underline{a}_3}{2}$. It is shown in Fig. 2.5(b). Metallic elements such as Cd, Zn, and Be crystallize in this structure.

1.4. The wurtzite structure

The wurtzite structure named after the French chemist C.A. Wurtz denotes the hexagonal phase of ZnS. This structure shown in Fig. 2.6 consists of two interpenetrating hcp lattices, where for one hcp the basis is formed by Zn at $\underline{0}$ and S at $\frac{\underline{a}_1}{3} + \frac{\underline{a}_2}{3} + \frac{\underline{a}_3}{2}$, while in the second hcp S is residing at $\underline{0}$ and Zn at $\frac{\underline{a}_1}{3} + \frac{\underline{a}_2}{3} - \frac{\underline{a}_3}{2}$. The separation between the two hcp lattices is the bond length between Zn and S. The structure exhibits a unique axis parallel to $\underline{a}_3$, the c-axis. The lattice constant in this direction is 2.67 times the bond length, while the in-plane lattice constant is 1.63 times the bond length. Since even in the wurtzite structure the atoms are tetrahedrally coordinated, this structure is sometimes referred to as similar to the zincblende structure. While in the zincblende structure (looking in the direction of the body diagonal of the cube) the adjacent planes of Zn and S atoms are rotated by 180° with respect to each other, this rotation is absent in the wurtzite structure. However, due to the completely different symmetry of the wurtzite structure, this comparison can be rather misleading. Although the number of nearest neighbors is four as for the zincblende structure, the number of atoms per primitive cell for the wurtzite structure increases to four in comparison to the zincblende structure. The wurtzite crystal structure is shown in Fig. 2.6.

2. Symmetry Properties of Crystal Lattices

In the previous section, we have focused on the translational symmetry of the crystal structure. Another possible classification can be achieved

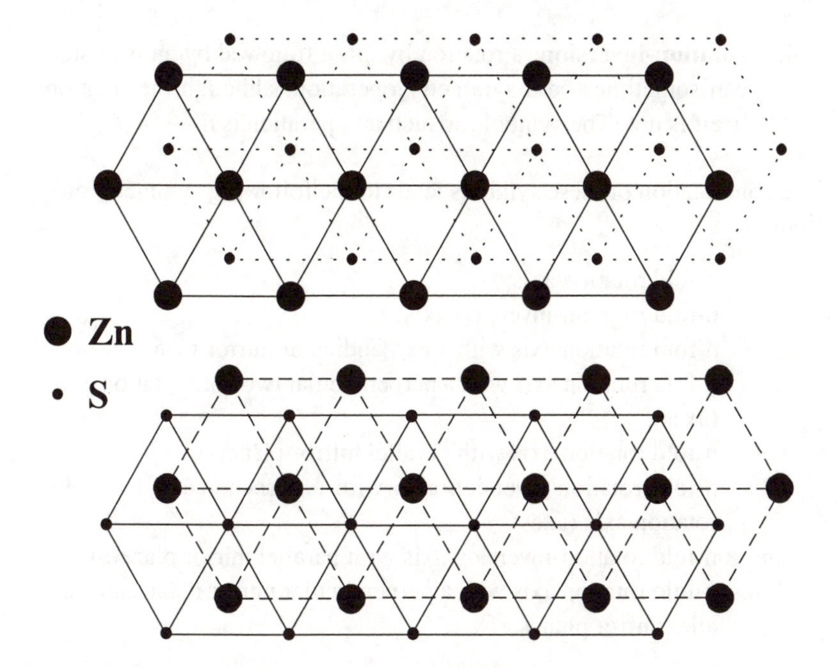

Fig. 2.6. The wurtzite crystal structure. The two planes with solid lines are stacked directly above each other. The plane with dotted lines is above the respective solid plane, while the plane with dashed lines is below the respective solid plane.

by considering symmetry operations, which map the crystal lattice into itself leaving at least one particular point of the lattice fixed. The set of operations for a particular crystal structure is called the point group of the crystal lattice. The possible operations are:

1. **rotations** through integral multiples of $2\pi/n$ about the same axis, where n=1, 2, 3, 4, or 6 refers to the n-fold axis. This operation is denoted by n,
2. **reflection** at a plane, e.g., $x \to -x$, $y \to y$, and $z \to z$, which is denoted by m,
3. **inversion** at a point, e.g., $x \to -x$, $y \to -y$, and $z \to -z$, which is denoted by $\bar{1}$,

4. **rotation-inversion**: a rotation by $2\pi/n$ followed by an inversion can sometimes be a symmetry operation, although the rotation itself is not. The symbol for such an operation is $\bar{n}$.

The combination of these symbols leads to the following symmetry operations:

n	n-fold rotation axis,
$\bar{n}$	n-fold rotation-inversion axis,
$\frac{n}{m}$	n-fold rotation axis with a perpendicular mirror plane,
$n2$	n-fold rotation axis with a perpendicular two-fold rotation axis (axes),
nm	n-fold rotation axis with parallel mirror plane(s),
$\bar{n}2$	n-fold rotation-inversion axis with a perpendicular two-fold rotation axis (axes),
$\bar{n}m$	n-fold rotation-inversion axis with parallel mirror plane(s),
$\frac{n}{m}m$	n-fold rotation axis with a perpendicular mirror plane and parallel mirror planes.

The diamond lattice belongs to the point group O_h, where O stands for octahedron and h for horizontal referring to an axis perpendicular to the rotational axis. In the above notation, this point group is written as $\frac{4}{m}\bar{3}\frac{2}{m}$. This implies that the diamond structure has a three-fold rotation-inversion axis, or equivalently a three-fold rotation axis and inversion symmetry, as well as a two-fold and four-fold rotation axis with mirror planes perpendicular to them. In comparison to the diamond structure, the zincblende structure has a lower symmetry. The inversion symmetry of the diamond structure is lost due to the two different atoms. The zincblende structure has the symmetry of the T_d point group, where T refers to tetrahedron and d to diagonal. It contains the operations $\bar{4}3m$, which include a four-fold rotation-inversion axis, a three-fold rotation axis, and mirror planes parallel to it. The three most important crystal structures for semiconductors together with their respective point group are listed in Tab. 2.1.

The wurtzite structure belongs to the C_{6v} point group, where C stands for cyclic, 6 for the six-fold rotational axis, and v for vertical, i.e., parallel to the rotational axis. The symmetry operations are $6mm$, which implies that the structure has a 6-fold rotation axis, a mirror plane containing the

Table 2.1. The classification of group IV, III-V, II-VI, and I-VII semiconductors according to their crystal structure, point, and space group. F refers to the face-centered and P to the primitive lattice.

Structure	Point group	Space group	Materials
diamond	$\frac{4}{m}\bar{3}\frac{2}{m}$ (O_h)	F$\frac{4_1}{d}\bar{3}\frac{2}{m}$ (O_h^7)	Diamond (C), Ge, Si, α-Sn
zincblende	$\bar{4}3m$ (T_d)	F$\bar{4}3m$ (T_d^2)	III-V: AlAs, AlP, AlSb, BN, BP, GaAs, GaP, GaSb, InAs, InP, InSb II-VI: CdS, CdSe, CdTe, β-HgS, HgSe, HgTe, ZnS, ZnSe, ZnTe I-VII: γ-CuBr, γ-CuCl, γ-CuI
wurtzite	$6mm$ (C_{6v})	P6_3mc (C_{6v}^4)	III-V: AlN, BN, GaN, InN II-VI: CdS, CdSe, ZnO, ZnS, ZnSe I-VII: β-AgI

axis of rotation, and as many additional mirror planes as the existence of the 6-fold axis requires.

If one adds translations as symmetry operations to the point group, one can obtain the respective space group of the crystal structure. The space group is not the crystal structure. However, if a crystal structure exhibits all the symmetry elements of the space group, the symmetry of the crystal structure is completely characterized. In addition to the point group operations and translations, the crystal structure may also have screw axes and glide-plane operations. A screw axis is a combination of a rotation and a translation parallel to the rotation axis. It is denoted by n_k $(k = 1, ..n-1)$, where k refers to the amount of translation in terms of the lattice constant in the respective direction divided by n. A glide plane combines a reflection plane and a translation parallel to the plane. In the diamond structure, there is a particular glide plane denoted as d. For the hexagonal structure, there is an axial glide plane denoted by c. Table 2.1 also lists the space groups for the three most important crystal structures for semiconductors. It also classifies most of the group IV, III-V, II-VI, and I-VII semiconductors according to their crystal structure.

3. The Reciprocal Lattice

The periodicity and symmetry of the lattice play a fundamental role in the determination of the structural and electronic properties. Crystal diffraction, the band structure, and the law of momentum conservation for crystal lattices depend strongly on the symmetry of the underlying crystal structure. All these properties are related to the momentum, i.e., the wavevector. For X-ray scattering, the change in the wavevector of the scattered X-ray waves is directly connected with the crystal momentum. The wavevector of an electron wave in a solid is also linked to the crystal momentum. It is therefore very useful to introduce a discrete Fourier-transformed lattice, the reciprocal lattice. The primitive vectors $\underline{b}_i$ of the reciprocal lattice are constructed from the primitive vectors $\underline{a}_i$ of the crystal lattice using the transformation

$$\underline{b}_i = 2\pi \frac{\underline{a}_j \times \underline{a}_k}{\underline{a}_i \cdot (\underline{a}_j \times \underline{a}_k)}, \tag{2.4}$$

where i, j, k are cyclically interchanged. The following orthogonality relation follows directly from Eq. (2.4) that

$$\underline{a}_i \cdot \underline{b}_j = 2\pi\, \delta_{ij}\,. \tag{2.5}$$

As for the direct lattice, every lattice point of the reciprocal lattice $\underline{K}$ can be written as a linear combination of these three primitive vectors $\underline{b}_i$, i.e.,

$$\underline{K} = \sum_{i=1}^{3} M_i\, \underline{b}_i\,, \tag{2.6}$$

where M_i denote integers. Equation (2.5) implies for every lattice point $\underline{R}$ that

$$\exp(i\,\underline{K}\cdot\underline{R}) = 1 \tag{2.7}$$

for all vectors $\underline{K}$ of the reciprocal lattice.

The reciprocal lattice of a Bravais lattice is again a Bravais lattice. Furthermore, the reciprocal lattice of the reciprocal lattice is the original direct lattice. In Tab. 2.2 the original and reciprocal lattices of the most important Bravais lattices are listed. One can easily construct the primitive vectors of the reciprocal lattice using Eq. (2.4) to proof the correspondence listed in this table. Note that, in the case of a lattice with a basis, one uses the reciprocal lattice determined by the underlying Bravais lattice. The basis is not taken into account. Therefore, the reciprocal lattice of the diamond and zincblende structure is also the body-centered cubic, while the reciprocal lattice of the wurtzite structure is a simple hexagonal. These structures are also included in Tab. 2.2.

The Wigner-Seitz cell of the reciprocal lattice is called the first Brillouin zone. Since we already know the Wigner-Seitz cells of the underlying Bravais lattices, we can directly deduce that the first Brillouin zone of the

Table 2.2. Reciprocal lattices and their lattice constants for the most common crystal structures. Sc denotes simple cubic, bcc body-centered cubic, fcc face-centered cubic, and sh simple hexagonal.

Crystal structure	Lattice constant	Reciprocal lattice	Lattice constant
sc	a	sc	$2\pi/a$
bcc	a	fcc	$4\pi/a$
fcc	a	bcc	$4\pi/a$
sh	a, c	sh	$4\pi/(\sqrt{3}a)$, $2\pi/c$
diamond, zincblende	a	bcc	$4\pi/a$
wurtzite	a, c	sh	$4\pi/(\sqrt{3}a)$, $2\pi/c$

Table 2.3. Wigner-Seitz cells of some crystal structures and their respective first Brillouin zones.

Crystal structure	Wigner-Seitz cell	1. Brillouin zone
sc	cube	cube
bcc	truncated octahedron	rhombic dodecahedron
fcc	rhombic dodecahedron	truncated octahedron
sh	hexagonal prism	hexagonal prism
diamond, zincblende		truncated octahedron
wurtzite		hexagonal prism

reciprocal lattice of the simple cubic lattice is a cube. The volume of the unit cell in reciprocal space is $(2\pi)^3/V$, where V denotes the volume of the unit cell in the direct lattice. This can easily be proven using Eq. (2.4). For the simple cubic lattice $V = a^3$. Therefore, the unit cell in reciprocal space has a value of $(2\pi/a)^3$. The Wigner-Seitz cells of some direct lattices and the corresponding first Brillouin zones are listed in Tab. 2.3.

4. Miller Indices and Points of High Symmetry in the First Brillouin Zone

There is a direct connection between a vector of the reciprocal lattice and crystal planes of the lattice structure. For example in the sc lattice, the vector describing the plane generated by $\underline{a}_1$ and $\underline{a}_2$ is perpendicular to $\underline{a}_1$ and $\underline{a}_2$, i.e., it is parallel to $\underline{b}_3$, a primitive vector of the reciprocal lattice. A convenient way to describe a plane is using the respective integer coefficients of the corresponding reciprocal lattice vector. In order to determine these, the vector product of the two vectors, which generate the plane, is formed and multiplied by $2\pi/V$, where V denotes the volume of the primitive cell of the lattice. This product can be expressed as a reciprocal lattice vector. The integers in front of the three primitive vectors hkl are known as the Miller indices. For example, for a plane, which is generated by the vectors $\underline{a}_1 + 2\underline{a}_3$ and $\underline{a}_1 + \underline{a}_2$, one gets -221. Negative numbers are indicated by placing a minus sign above the respective index, i.e., $\bar{2}21$. A face of a cube is denoted by 100, a plane along the body diagonal of the cube by 111. Different brackets are used for denoting single planes, equivalent planes, directions, and equivalent directions, which are listed below.

(hkl)	a single plane or a set of parallel planes,
$\{hkl\}$	set of planes of equivalent symmetry, e.g., for cubic symmetry $\{100\}$ refers to (100), (010), (001), $(\bar{1}00)$, $(0\bar{1}0)$, and $(00\bar{1})$,
$[hkl]$	a crystalline direction, e.g., [100] refers to the x-axis,
$\langle hkl \rangle$	full set of equivalent directions,
$[a_1 a_2 a_3 c]$	notation for hexagonal lattices, the c axis is [0001].

a_1, a_2, and a_3 correspond to the integer coefficients of three vectors in the hexagonal plane with a relative angle of 120° fulfilling the relation $a_1 + a_2 + a_3 = 0$. The position of points in a unit cell are specified in terms of the lattice coordinates, in which each coordinate is a fraction of the axial length a, b, or c in the direction of the coordinate. For the Wigner-Seitz cell, the origin is at the center of the cell.

The first Brillouin zone contains a number of points of higher symmetry, which are particularly useful in describing the band structure. In Tab. 2.4 some of these points are listed for the two most important semiconductor lattices, the fcc and the hexagonal lattice. Γ denotes the center of the Brillouin zone, while other points with lower symmetry are also indicated. The first Brillouin zone of both lattices together with the points of higher symmetry are shown in Fig. 2.7.

Table 2.4. Coordinates of points of higher symmetry in the first Brillouin zones of the fcc and hexagonal lattices.

Point	Γ	Δ	Λ	Σ	L	W	X
fcc	000	$0k_y0$	$k_xk_yk_z$	k_xk_y0	$\frac{1}{4}\frac{1}{4}\frac{1}{4}$	$\frac{1}{4}\frac{1}{2}0$	$0\frac{1}{2}0$

Point	Γ	Δ	Λ	Σ	A	K	M
hexagonal	000	$00k_z$	k_1k_20	k_100	$00\frac{1}{2}$	$\frac{1}{3}\frac{1}{3}0$	$\frac{1}{2}00$

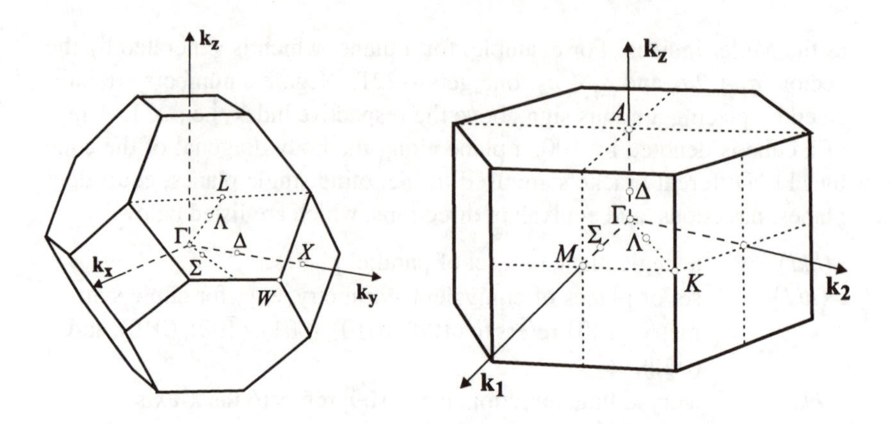

Fig. 2.7. The symmetry points of the first Brillouin zone for the fcc (left) and hexagonal lattice (right).

The group IV semiconductors and a number of III-V semiconductors in the zincblende structure exhibit preferential cleavage planes. In the diamond-type crystals, cleavage occurs along the {111} planes. However, in the zincblende structure, the layers of atoms forming {111} planes are alternately composed of only group III and group V atoms (cf. Fig. 2.8). If these atoms are differently charged, the electrostatic attraction between these planes will make it difficult to separate them. The (110) planes, how-

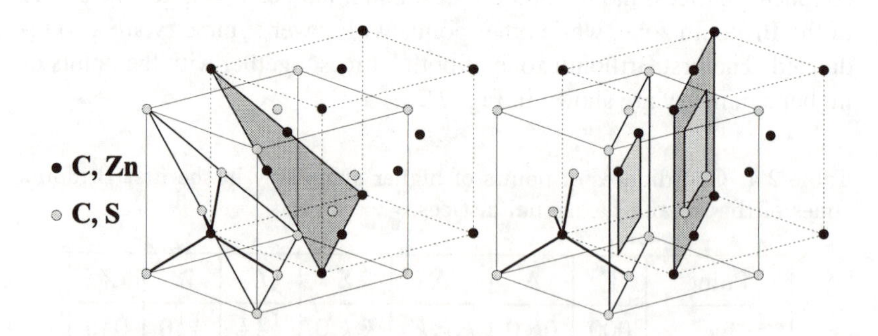

Fig. 2.8. The shaded areas indicate the dominant cleavage planes in the diamond (left) and zincblende (right) structures, which are (111) and (110), respectively.

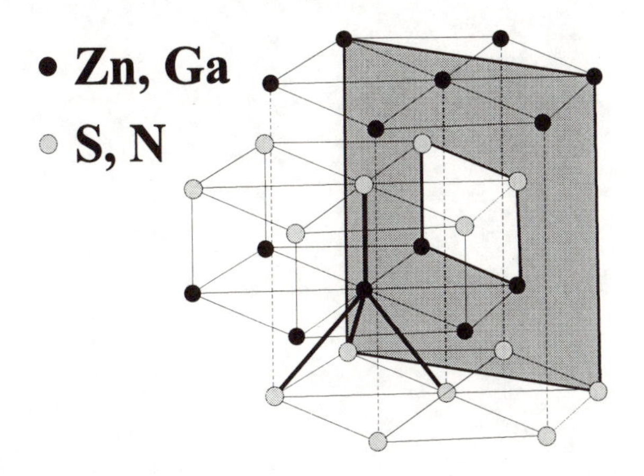

Fig. 2.9. The shaded areas indicate two examples of cleavage planes for the wurtzite structure denoted $[10\bar{1}0]$ and $[\bar{1}\bar{1}20]$.

ever, are each composed of equal numbers of group III and group V atoms (cf. Fig. 2.8). Therefore, there is no overall electrostatic force between them. If the zincblende crystal exhibits some degree of ionicity, the cleavage along (110) planes will be preferred. For example, GaP cleaves only along (110). AlSb and GaAs also show additional weak cleavage along (111) planes. GaSb, InAs, and InSb exhibit in addition to cleavage along (110) planes an even larger degree of cleavage along (111) planes. For the wurtzite structure, there are two sets of cleavage planes, which are perpendicular to the hexagonal plane. Each set consists of three planes with an angle of 120° between each other. The first set is parallel to the hexagonal faces, i.e., it can be described by $[10\bar{1}0]$ and two equivalent vectors. The second set is rotated by 60° with respect to the first set, i.e., it can be described by $[2\bar{1}\bar{1}0]$ and two equivalent vectors. One example of each set of cleavage planes for the wurtzite structure is shown in Fig. 2.9.

CHAPTER 3
ELECTRONS IN A PERIODIC POTENTIAL

In semiconductors, the formation of energy bands and energy gaps is closely related to the existence of a periodic potential with the same symmetry properties as the crystal lattice. Although the formation of energy bands alone can be understood from the interaction of a very large number of identical atoms with degenerate energy levels, the existence of energy gaps is more difficult to deduce from this chemical approach. In this chapter, the effect of the periodic potential on the energy states and wave functions will be discussed. We will pay particular attention to the formation of energy gaps, which are essential for the semiconducting properties.

There are two approaches, which describe the appearance of energy gaps. The first one treats the periodic potential as a perturbation. It is usually referred to as the nearly free electron model. This model is actually more applicable to metals. The second approach, which is called the tight-binding model, takes into account that the electrons are strongly bonded to the individual atoms. It is more suitable for semiconductors and insulators. We will focus on the latter approach. Most properties of semiconductors are determined by the band structure in the vicinity of the energy gap. There is a semiempirical model called the $\underline{k} \cdot \underline{p}$ model, which describes the band structure of semiconductors quite accurately near the energy gap. This method will be discussed in the next chapter. Here we will focus on some general properties of energy states and wave functions in a periodic potential.

1. The Bloch Theorem

The periodicity of the crystal structure results in a periodic potential $V(\underline{r})$

$$V(\underline{r}+\underline{R}) = V(\underline{r}) , \tag{3.1}$$

which is invariant under translations by a lattice vector $\underline{R}$ as defined in Eq. 2.1. $V(\underline{r})$ determines the energy states and wave functions of the electrons. Each function on the atomic scale, which is related to a physical quantity, must possess this periodicity. In particular, the electronic wave function $\Psi(\underline{r})$ should be such that a translation by a lattice vector does not change the probability density, i.e., $|\Psi(\underline{r})|^2$. Therefore,

$$\Psi(\underline{r}+\underline{R}) = \Psi(\underline{r})\, e^{i\phi(\underline{R})} , \tag{3.2}$$

where $\phi(\underline{R})$ denotes a phase, which depends on the lattice vector $\underline{R}$. The Hamiltonian $H(\underline{r})$ of the solid is the sum of the kinetic energy $T(\underline{r})$ and the potential energy $V(\underline{r})$. Since the kinetic term is invariant under any translation, the Hamiltonian is invariant under any translation within the crystal structure, i.e., $H(\underline{r}+\underline{R}) = H(\underline{r})$. Defining an operator for lattice translations $T_{\underline{R}}$, it can easily be shown that

$$\begin{aligned} T_{\underline{R}}[H(\underline{r})\,\Psi(\underline{r})] &= H(\underline{r}+\underline{R})\,\Psi(\underline{r}+\underline{R}) \\ &= H(\underline{r})\,[T_{\underline{R}}\,\Psi(\underline{r})] . \end{aligned} \tag{3.3}$$

Therefore, the Hamiltonian and the lattice translation operator commute so that the eigenfunctions of the Hamiltonian are also eigenfunctions of $T_{\underline{R}}$. If one combines two translations, it follows that

$$\begin{aligned} T_{\underline{R}}\,T_{\underline{R}'}\,\Psi(\underline{r}) &= T_{\underline{R}}\,\Psi(\underline{r}+\underline{R}') = \Psi(\underline{r}+\underline{R}'+\underline{R}) \\ &= T_{\underline{R}'+\underline{R}}\,\Psi(\underline{r}) = T_{\underline{R}'}\,T_{\underline{R}}\,\Psi(\underline{r}) . \end{aligned} \tag{3.4}$$

The eigenvalues $c(\underline{R})$ of $T_{\underline{R}}$ therefore obey the following relation

$$c(\underline{R})\,c(\underline{R}') = c(\underline{R}+\underline{R}') . \tag{3.5}$$

Since we already know that the effect of a lattice translation is to change the wave function by a phase factor, i.e., $c(\underline{R}) = e^{i\phi(\underline{R})}$, we conclude from Eq. 3.5 that the phase factor has to obey the following relation

$$\phi(\underline{R}) + \phi(\underline{R}') = \phi(\underline{R}+\underline{R}') . \tag{3.6}$$

ϕ is therefore a linear function of $\underline{R}$, which can be expressed as

$$\phi(\underline{R}) = \underline{k}\cdot\underline{R} . \tag{3.7}$$

$\underline{k}$ is an arbitrary vector in reciprocal space, i.e., it can be written as

$$\underline{k} \;=\; \sum_{i=1}^{3} y_i \, \underline{b}_i \,, \tag{3.8}$$

where the $\underline{b}_i$ denote the primitive vectors of the reciprocal lattice and the y_i are arbitrary numbers. If the y_i are integers, $\underline{k}$ is a reciprocal lattice vector. To summarize, the Bloch theorem states the following: The eigenfunctions of a Hamiltonian $H(\underline{r})$ with $V(\underline{r}+\underline{R}) = V(\underline{r})$ can be chosen as

$$\Psi_{\underline{k}}(\underline{r}) \;=\; u_{\underline{k}}(\underline{r}) \, e^{i\underline{k}\cdot\underline{r}} \,, \tag{3.9}$$

where

$$u_{\underline{k}}(\underline{r}+\underline{R}) \;=\; u_{\underline{k}}(\underline{r}) \tag{3.10}$$

are the Bloch functions, which exhibit the same translational symmetry as the crystal structure. $\underline{k}$ denotes the crystal (sometimes also quasi- or pseudo-) wavevector of the electron moving in the crystal potential. It differs from the normal momentum, since we can always add a reciprocal lattice vector without changing the probability density. By choosing an appropriate boundary condition, the number of allowed values for $\underline{k}$ will be restricted. Usually, when bulk properties are described, periodic boundary conditions are selected

$$\Psi(\underline{r}+N_i\underline{a}_i) \;=\; \Psi(\underline{r}) \quad \text{for} \quad i = 1,2,3 \,, \tag{3.11}$$

where $\underline{a}_i$ are the three primitive vectors of the lattice and N_i are integers of the order of $N^{1/3}$ with $N = N_1N_2N_3$ denoting the total number of primitive cells in the crystal. Applying Bloch's theorem leads to

$$e^{iN_i\underline{k}\cdot\underline{a}_i} \;=\; 1 \quad \text{for} \quad i = 1,2,3 \,. \tag{3.12}$$

Using Eq. 3.8 for $\underline{k}$ results in the condition that

$$2\pi \, N_i \, y_i \;=\; 2\pi \, M_i \,, \tag{3.13}$$

where M_i are integers. Therefore, the allowed Bloch wavevectors can be written as

$$\underline{k} \;=\; \sum_{i=1}^{3} \frac{M_i}{N_i} \, \underline{b}_i \,. \tag{3.14}$$

It is convenient to restrict the values of $\underline{k}$ to the first Brillouin zone, which completely describes the crystal structure. The wave functions and energy levels $E_{\underline{k}}$ are periodic functions of $\underline{k}$ in the reciprocal lattice according to

$$\begin{aligned} \Psi_{\underline{k}+\underline{K}}(\underline{r}) &= \Psi_{\underline{k}}(\underline{r}) \\ E_{\underline{k}+\underline{K}} &= E_{\underline{k}} . \end{aligned} \tag{3.15}$$

The Schrödinger equation for the total wave function

$$\left[-\frac{\hbar^2}{2m}\underline{\nabla}^2 + V(\underline{r})\right] \Psi_{\underline{k}}(\underline{r}) = E_{\underline{k}}\, \Psi_{\underline{k}}(\underline{r}) \tag{3.16}$$

can be converted into the Schrödinger equation for the Bloch functions using Eq. 3.9. The effect of the momentum operator $\underline{p} = \frac{\hbar}{i}\underline{\nabla}$ on the full wave function is

$$\underline{p}\, \Psi_{\underline{k}}(\underline{r}) = \exp(i\underline{k}\cdot\underline{r})\, \left(\underline{p} + \hbar\underline{k}\right) u_{\underline{k}}(\underline{r}) . \tag{3.17}$$

Repeating the operation of $\underline{p}$ again leads to

$$\underline{p}^2\, \Psi_{\underline{k}}(\underline{r}) = \exp(i\underline{k}\cdot\underline{r})\, \left(\underline{p}^2 + 2\,\hbar\, \underline{k}\cdot\underline{p} + \hbar^2\underline{k}^2\right) u_{\underline{k}}(\underline{r}) . \tag{3.18}$$

We therefore obtain the following Schrödinger equation for the Bloch functions $u_{\underline{k}}(\underline{r})$

$$\begin{aligned} \left[-\frac{\hbar^2}{2m}\underline{\nabla}^2 - i\,\frac{\hbar^2}{m}\underline{k}\cdot\underline{\nabla} + V(\underline{r})\right] & u_{\underline{k}}(\underline{r}) = \\ \left[E_{\underline{k}} - \frac{\hbar^2\underline{k}^2}{2m}\right] & u_{\underline{k}}(\underline{r}) . \end{aligned} \tag{3.19}$$

This equation is the basis for the $\underline{k}\cdot\underline{p}$ method, which will be described in the next chapter.

2. The Kronig-Penney Model

The solution of the Schrödinger equation for a periodic one-dimensional potential was first obtained by Kronig and Penney in 1931. In their original publication, they used a one-dimensional periodic potential with two different values for the potential, which is a fair approximation for a covalent solid. We will use the rectangular periodic potential shown in Fig. 3.1. The periodic potential $V(z+nd) = V(z)$ of Fig. 3.1 is given by

$$V(z) = \begin{cases} 0 & \text{for} \quad 0 < z - nd \leq a \\ V_0 & \text{for} \quad a < z - nd \leq b \end{cases} . \tag{3.20}$$

Since the potential is constant in both regions, the solution of the Schrödinger equation for $E < V_0$ is given by

$$\Psi(z) = \begin{cases} A_1 \, e^{ik_z z} + A_2 \, e^{-ik_z z} & \text{for} \quad 0 < z - nd \leq a \\ B_1 \, e^{\kappa_z z} + B_2 \, e^{-\kappa_z z} & \text{for} \quad a < z - nd \leq b \end{cases} , \tag{3.21}$$

where $\hbar^2 k_z^2 = 2mE_{k_z}$ and $\hbar^2 \kappa_z^2 = 2m(V_0 - E_{k_z})$. The boundary condition at $z = 0$ for the continuity of the wave function $\Psi(0^+) = \Psi(0^-)$, where 0^+ denotes the well region and 0^- the barrier region, and its derivative $\Psi'(0^+) = \Psi'(0^-)$, where $'$ denotes the first derivative $\Psi' = d\Psi/dz$, results in two equations for the coefficients A_1, A_2, B_1, and B_2

$$A_1 + A_2 = B_1 + B_2 \quad \text{and} \tag{3.22}$$

$$ik_z (A_1 - A_2) = \kappa_z (B_1 - B_2) . \tag{3.23}$$

Two additional equations are obtained, when the periodicity of the wave function and its derivative are taken into account

$$\begin{aligned} \Psi_k(z-d) &= \Psi_k(z) \, e^{-ikd} \quad \text{and} \\ \left. \frac{d\Psi_k(z)}{dz} \right|_{z-d} &= \left. \frac{d\Psi_k(z)}{dz} \right|_z e^{-ikd} . \end{aligned} \tag{3.24}$$

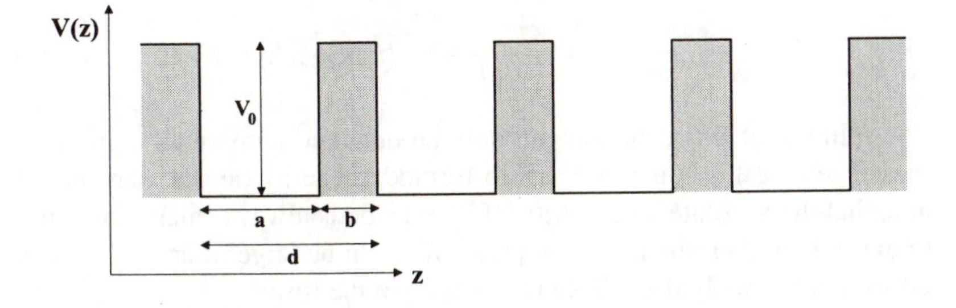

Fig. 3.1. The rectangular periodic potential used in the Kronig-Penney model.

Writing down these conditions for $z = a$ leads to two additional equations for the coefficients.

$$(A_1 e^{ik_z a} + A_2 e^{-ik_z a}) e^{-ikd} = B_1 e^{-\kappa_z b} + B_2 e^{\kappa_z b} , \tag{3.25}$$

$$\left(A_1 e^{ik_z a} - A_2 e^{-ik_z a}\right) e^{-ikd} = \frac{\kappa_z}{ik_z} \left(B_1 e^{-\kappa_z b} - B_2 e^{\kappa_z b}\right) . \tag{3.26}$$

The four equations 3.22, 3.23, 3.25, and 3.26 form a set of homogeneous equations, which can be written in matrix form

$$\underline{\underline{M}} \begin{pmatrix} A_1 \\ A_2 \\ B_1 \\ B_2 \end{pmatrix} = 0 , \tag{3.27}$$

where the matrix $\underline{\underline{M}}$ reads

$$\underline{\underline{M}} = \begin{pmatrix} 1 & 1 & -1 & -1 \\ e^{i(k_z a - kd)} & e^{-i(k_z a + kd)} & -e^{-\kappa_z b} & -e^{\kappa_z b} \\ ik_z & -ik_z & -\kappa_z & \kappa_z \\ ik_z \, e^{i(k_z a - kd)} & -ik_z \, e^{-i(k_z a + kd)} & -\kappa_z \, e^{-\kappa_z b} & \kappa_z \, e^{\kappa_z b} \end{pmatrix} . \tag{3.28}$$

In order to solve this set of equations, the determinant of the matrix in Eq. 3.28 has to vanish. This condition results in the following relationship

$$\begin{aligned} \cos(k\,d) &= \cos(k_z\,a)\cosh(\kappa_z\,b) \\ &\quad + \frac{\kappa_z^2 - k_z^2}{2\kappa_z k_z} \sin(k_z\,a)\sinh(\kappa_z\,b) . \end{aligned} \tag{3.29}$$

The solution of this equation can only be obtained numerically. In order to realize that this equation leads to forbidden energy ranges, one should note that the absolute value of its left-hand side is always smaller or equal to unity. However, the right-hand side (rhs) can be larger than 1. This can be seen more easily if Eq. 3.29 is rewritten in the form

$$\cos(k\,d) = \frac{\cos(k_z\,a + \delta)}{T} \tag{3.30}$$

with

$$T = \frac{\cos(\delta)}{\cosh(\kappa_z b)} \quad \text{and}$$

$$\tan(\delta) = \frac{k_z^2 - \kappa_z^2}{2 k_z \kappa_z} \tanh(\kappa_z b) . \tag{3.31}$$

The denominator T in Eq. 3.30 will always be smaller than unity for some energy intervals. The rhs of Eq. 3.29 is plotted in Fig. 3.2 vs energy. The parameters have been selected to obtain several well-defined bands of allowed states (gray regions) with a significant band width. The widths of the energy bands increases with increasing energy, while the width of the gaps (white regions) remains almost constant. This behavior, however, depends on the actual form of the potential used. A number of energy bands and gaps is present. In the following, we will label the energy bands by $E_n(k)$ with $n = 1,2,3....$ The energy dispersion is obtained by plotting the energy versus the crystal wavevector k. For narrow energy bands sepa-

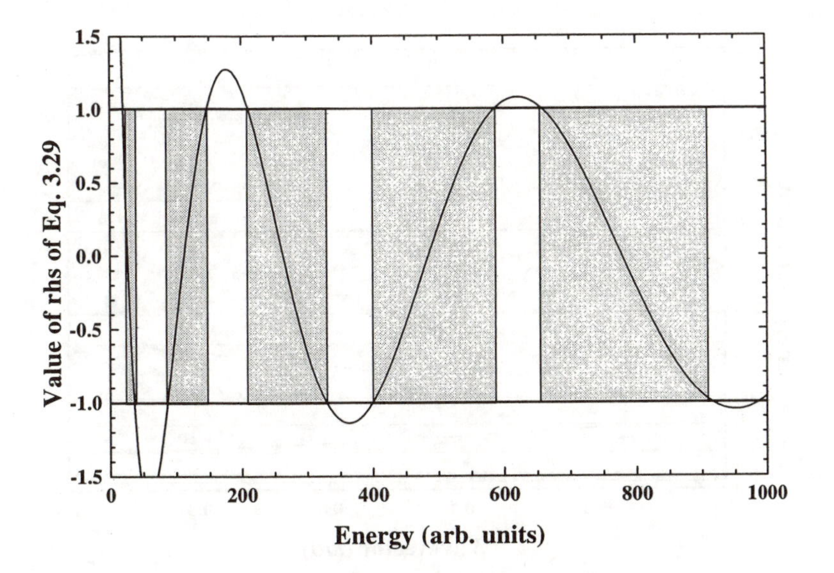

Fig. 3.2. Allowed (gray) and forbidden (white) energy regions for the one-dimensional periodic potential shown in Fig. 3.1 for $E < V_0$.

rated by wide gaps, i.e., in the tight-binding limit, the dispersion can be approximated by

$$E_n(k) = E_n^c + (-1)^n \frac{\Delta_n}{2} \cos(k\,d)\,, \tag{3.32}$$

where E_n^c denotes the center of the energy bands and Δ_n their widths. $k\,d$ runs from $-\pi$ to π. E_n^c is obtained for $k\,d = \pi/2$. In the tight-binding limit, the band width can be approximated by

$$\Delta_n = \frac{8(V_0 - E_n^c)}{\kappa(E_n^c)\, F'[\kappa(E_n^c)]} \exp(-\kappa(E_n^c)\, b)\,, \tag{3.33}$$

where F' denotes the first derivative of the rhs of Eq. 3.29 with respect to κ. The band width decreases exponentially with increasing barrier width in this limit. Figure 3.3 shows the energy dispersion for the same parameters used for Fig. 3.2. Due to the symmetry of the energy dispersion with respect to changing k to $-k$ (cf. Eq. 3.32), the energy dispersion is only

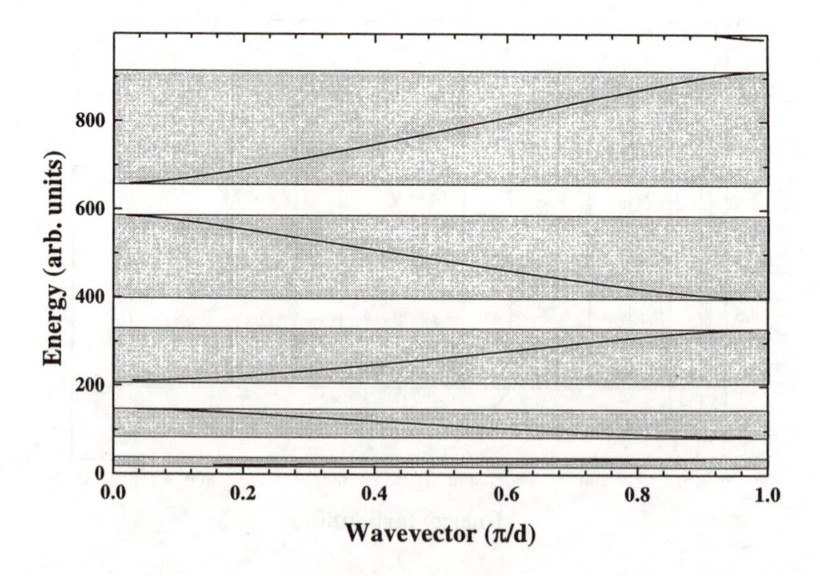

Fig. 3.3. Energy dispersion for the same parameters used in Fig. 3.2.

shown for positive values of k. The energy bands have approximately the tight-binding form of Eq. 3.32.

If the energy E is larger than the barrier height V_0, we have to set κ_z in Eq. 3.21 equal to iK_z so that the solution of the Schrödinger equation becomes

$$\Psi_{k_z}(z) = \begin{cases} A_1\, e^{ik_z z} + A_2\, e^{-ik_z z} & \text{for} \quad 0 < z - nd \leq a \\ B_1\, e^{iK_z z} + B_2\, e^{-iK_z z} & \text{for} \quad a < z - nd \leq b \end{cases}, \qquad (3.34)$$

where $\hbar^2 K_z^2 = 2m(E - V_0)$. The equation determining the energy levels reads

$$\begin{aligned} \cos(k\, d) &= \cos(k_z\, a) \cos(K_z\, b) \\ &\quad - \frac{K_z^2 + k_z^2}{2\, K_z k_z} \sin(k_z\, a) \sin(K_z\, b)\,. \end{aligned} \qquad (3.35)$$

Eq. 3.35 can also be rewritten in the form of Eq. 3.29 with

$$\begin{aligned} T &= \frac{\cos(\delta)}{\cos(K_z\, b)} \quad \text{and} \\ \tan(\delta) &= \frac{k_z^2 + K_z^2}{2\, k_z\, K_z} \tan(K_z\, b)\,. \end{aligned} \qquad (3.36)$$

There are still energy values, for which T is larger than one so that energy gaps will occur. The value of the rhs of Eq. 3.35 is shown in Fig. 3.4, where allowed (gray) and forbidden (white) energy regions are present. However, the energy gaps are now much smaller than the energy bands compared to Fig. 3.2. The parameters used in generating Fig. 3.4 are slightly different from the ones for Fig. 3.2. When $E \gg V_0$, the nearly free electron model is recovered, in which the bands are very broad, while the gaps are very small.

The Bloch wave functions for this one-dimensional problem have the form

$$\begin{matrix} A_1\, e^{i(k_z - k)z} + A_2\, e^{-i(k_z + k)z} & \text{for} \quad 0 < z - nd \leq a \\ B_1\, e^{(\kappa_z - ik)z} + B_2\, e^{-(\kappa_z + ik)z} & \text{for} \quad a < z - nd \leq b \end{matrix}\,. \qquad (3.37)$$

These equations do not look particularly illuminating. Nevertheless, the concept of the Bloch wave functions is very important as shown in this chapter, since these functions exhibit the symmetry of the crystal structure.

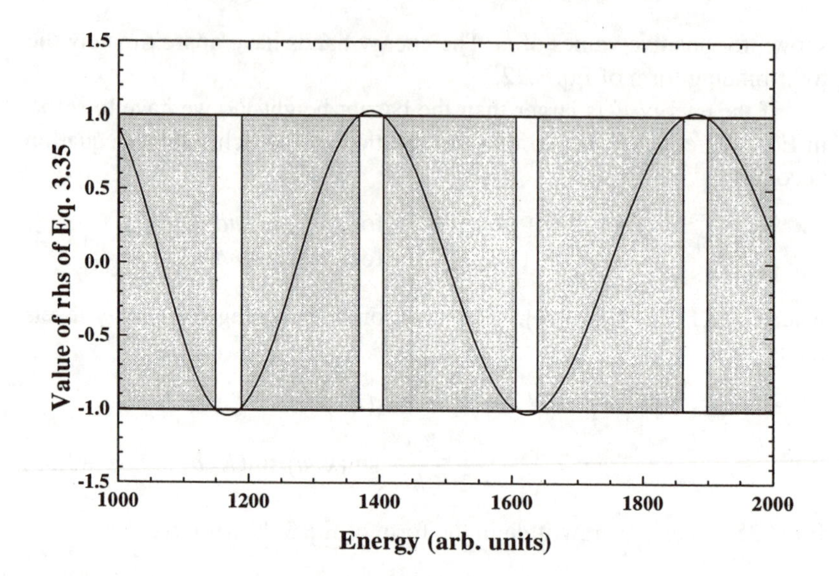

Fig. 3.4. Allowed (gray) and forbidden (white) energy regions for the one-dimensional periodic potential shown in Fig. 3.1 for $E > V_0$.

In three dimensions, energy bands and gaps can also exist. However, the simultaneous appearance of an energy gap in all three dimensions does not always occur, e.g., semimetals have a small negative energy gap. Furthermore, metals usually do not have a three-dimensional energy gap at all. The presence of energy gaps is strongly correlated with the type of bonding in the solid. Covalent bonding leads to small and wide energy gaps (semiconductors and insulators), ionic bonding to wide energy gaps (insulators), and metallic bonding to no energy gaps at all (metals). To determine the actual band structure of a semiconductor, one therefore has to consider the crystal structure together with the bonding between the different atoms.

CHAPTER 4

MODELS OF BAND STRUCTURE: ELECTRONS AND HOLES

The periodic potential and its symmetry determine the energy band structure of the semiconductor. Since the outer shell in covalently bonded semiconductors is completely filled with electrons obeying Pauli's principle, the energy band just below the energy gap is completely filled. This highest occupied band is called the valence band, while the first unoccupied band just above the energy gap is denoted as the conduction band. These are the most important bands to consider for semiconductors. Many electrical and optical properties of semiconductors are determined by the band structure in the vicinity of the energy gap, i.e., at the top of the valence and at the bottom of the conduction band. There is a very powerful abstraction, the effective mass concept, which puts all the information of the band structure near the top of the valence band or bottom of the conduction band into one quantity, the respective effective mass.

1. The Concept of the Effective Mass

The energy of a free electron is given by

$$E_n(\underline{k}) = \frac{\hbar^2 \underline{k}^2}{2m} . \tag{4.1}$$

The energy of an electron in a solid near a maximum or minimum of the band structure can be expanded in a power series in $\underline{k}$. Since the first derivative vanishes near a maximum or minimum of the band structure, the leading term will be of order $\underline{k}^2$. In this sense, charge carriers in a solid can be described as free electrons with an effective mass, which is determined by the curvature of the band structure in the vicinity of the energy gap. Using the one-dimensional tight-binding formula of Eq. 3.32, the energy near a minimum or maximum can be written as

$$E_n(k) \;=\; E_n^c + (-1)^n \, \frac{\Delta_n}{2} \left(1 - \frac{k^2 \, d^2}{2} + \cdots \right). \qquad (4.2)$$

Rearranging the terms and combining the factors in front of k^2 results in

$$E_n(k) \;=\; E_n^0 + (-1)^{n+1} \, \frac{\hbar^2 \, k^2}{2 \, m_n^*} + \cdots, \qquad (4.3)$$

where $E_n^0 = E_n^c + (-1)^n \, \Delta_n/2$ denotes the top (n even) or bottom (n odd)of the respective band and

$$m_n^* \;=\; \frac{2 \, \hbar^2}{\Delta_n \, d^2} \qquad (4.4)$$

denotes the effective mass. In this limit, the effective mass is inversely proportional to the energy band width. For a general band structure, one can determine the effective mass by taking the second derivative of the energy with respect to k at the maximum or minimum of the respective band, i.e.,

$$\frac{1}{m_n^*} \;=\; \frac{1}{\hbar^2} \, \frac{d^2 E_n(k)}{dk^2} \Big|_{\frac{dE_n(k)}{dk}=0} . \qquad (4.5)$$

In three dimensions, the effective masses can vary between different directions. The effective mass becomes an effective mass tensor defined through

$$\frac{1}{\underline{\underline{m_n^*}}} = \left(\frac{1}{m_n^*} \right)_{ij} \;=\; \frac{1}{\hbar^2} \, \frac{\partial^2 E_n(\underline{k})}{\partial k_i \partial k_j} \Big|_{\nabla_{\underline{k}} E_n(\underline{k}) = \underline{0}} . \qquad (4.6)$$

The effective mass tensor is a symmetric tensor, which can always be transformed into a diagonal form. However, the three diagonal elements do not necessarily have the same value. Terms of higher than second order in $\underline{k}$ can be included in the effective mass, which then becomes wavevector dependent. Even in the parabolic approximation given in Eq. 4.3, the effective mass can become energy dependent. It is usually referred to as the non-parabolicity of the effective mass.

2. Electrons and Holes

In semiconductors and insulators at zero temperature, the valence band is completely filled and the conduction band completely empty. Since in an empty as well as in a full band the electrons cannot carry any current, a semiconductor behaves at zero temperature like an insulator. In order to achieve electrical conduction, electrons have to be excited into the conduction band, e.g., by thermal excitation across the energy gap or by doping the semiconductor. Doping can be achieved by replacing one of the atoms of the semiconductor with another atom, which has one electron more or less in the outer shell than the constituent atoms. In order to describe the transport of electrons within an energy band, we have to determine the group velocity $\underline{v}_G$ of the electron wave packet

$$\begin{aligned} \underline{v}_G &= \nabla_{\underline{k}}\, \omega(\underline{k}) \\ &= \frac{1}{\hbar}\, \nabla_{\underline{k}} E(\underline{k})\,. \end{aligned} \tag{4.7}$$

For a free electron, the group velocity $\underline{v}_G$ is the same as the particle velocity $\underline{v}$, i.e., $\underline{v}_G = \underline{v} = \hbar\underline{k}/m = \underline{p}/m$.

The dynamics of an electron in an energy band is described by

$$\begin{aligned} \left(\frac{d\mathrm{v}_G}{dt}\right)_i &= \frac{1}{\hbar}\frac{d}{dt}\left(\nabla_{\underline{k}}\, E\right)_i \\ &= \frac{1}{\hbar}\sum_{j=1}^{3}\frac{\partial^2 E(\underline{k})}{\partial k_i\, \partial k_j}\,\frac{dk_j}{dt}\,. \end{aligned} \tag{4.8}$$

Using the definition of the effective mass tensor from Eq. 4.6, this equation can be written in a more compact form

$$\begin{aligned} \frac{d\underline{v}_G}{dt} &= \frac{1}{\underline{\underline{m^*}}} \cdot \frac{d\,\hbar\underline{k}}{dt} \\ &= \frac{1}{\underline{\underline{m^*}}} \cdot \frac{d\underline{p}}{dt}\,. \end{aligned} \tag{4.9}$$

The equation of motion for a particle experiencing a force $\underline{F}$ is

$$\frac{d\underline{p}}{dt} = \underline{F}\,. \tag{4.10}$$

Therefore, the acceleration of an electron inside an energy band is given by

$$\frac{d\underline{v}_G}{dt} = \frac{1}{\underline{\underline{m^*}}} \cdot \underline{F} . \tag{4.11}$$

For a free electron, the corresponding equation reads

$$\frac{d\underline{v}}{dt} = \frac{1}{m} \underline{F} . \tag{4.12}$$

Comparing this equation with Eq. 4.11 leads to the following conclusion regarding the description of carrier motion in an energy band: all the details of the band structure can be put in the effective mass tensor. The electrons behave as if they would be free particles with an effective mass m^*. This is a very powerful concept, which results in a great simplification of the description of transport properties in semiconductors. If the electron experiences an external electric field $\underline{F}_E$, Eq. 4.11 becomes

$$\frac{d\underline{v}_G}{dt} = -e \frac{1}{\underline{\underline{m^*}}} \cdot \underline{F}_E , \tag{4.13}$$

where e denotes the elementary charge. At the bottom of the conduction band, the effective mass is positive, since the energy increases in all directions of $\underline{k}$. However, at the top of the valence band, the effective mass becomes negative. Therefore, the motion of an electron at the top of the valence band can alternatively be described as that of a particle with a positive effective mass and a positive charge. These particles are called holes. The hole corresponds to a missing electron in the valence band. Since at zero temperature the valence band is fully occupied, all hole states are empty. The description of carrier transport in the valence band through holes is therefore equivalent to the description of the carrier transport in the conduction band through electrons. The energy of holes is usually taken to be zero at the top of the valence band and increases into the valence band. Since the velocity of the electron and the corresponding hole are the same, the crystal wavevector of the electron and that of the corresponding hole have to be opposite (cf. Eq. 4.9). The following correspondences between electron and hole properties exist:

1. The effective mass of the hole is $m_h^* = -m_e^*$.
2. The charge of the hole is $+e$, i.e., the opposite of the electron charge.
3. The velocity of the hole is equal to the velocity of the corresponding electron.
4. The energy of the hole is $E_h = -E_e$, where E_e denotes the electron energy.
5. The crystal wavevector of the hole is $\underline{k}_h = -\underline{k}_e$ of the electron.

The effective mass concept has also very unusual consequences. Returning to the tight-binding expression for the energy dispersion in one dimension, the effective mass becomes infinite at the inflection point, i.e., at $k\,d = \pi/2$. This implies that the carrier acceleration becomes zero at the inflection point. For larger values of $k\,d > \pi/2$, it changes sign, i.e., the carrier is accelerated in the opposite direction (hole regime) than for $k\,d < \pi/2$.

A very simple application of this concept can be carried out for the energy dispersion of Eq. 3.32. For the lowest energy band ($n = 1$), the group velocity is given by

$$v_G(k) = \frac{\Delta_1\, d}{2\,\hbar} \sin(k\,d)\,. \tag{4.14}$$

If a constant electric field is applied, the time evolution of the crystal wavevector follows $k(t) = -eFt/\,\hbar$. Therefore, the velocity of the electron oscillates in time according to

$$v_G(t) = \frac{\Delta_1\, d}{2\,\hbar} \sin(\frac{-edF}{\hbar}\, t)\,. \tag{4.15}$$

The frequency $\omega = 2\pi\nu$ of these oscillations is known as the Bloch or Zener frequency

$$\nu = \frac{e\,d\,F}{2\pi\,\hbar}\,. \tag{4.16}$$

For a typical lattice constant of 0.5 nm and an electric field of 100 $\mathrm{kVcm^{-1}}$, the Bloch frequency is 1.2 THz or the oscillation period is 0.83 ps. These Bloch or Zener oscillations have been observed only a few years ago in semiconductor superlattices, which exhibit such a tight-binding band structure.

3. The $\underline{k} \cdot \underline{p}$ Model

The $\underline{k} \cdot \underline{p}$ method is a semiempirical method to calculate the band structure using experimentally obtained quantities. The basis of this method is Eq. 3.19 for the Bloch wave function $u_{n,\underline{k}}(\underline{r})$ of the nth energy band. For $k = 0$, Eq. 3.19 becomes particularly simple

$$\left[-\frac{\hbar^2}{2m}\,\nabla^2 + V(\underline{r})\right] u_{n,\underline{0}}(\underline{r}) \;=\; E_{n,\underline{0}}\, u_{n,\underline{0}}(\underline{r})\,. \tag{4.17}$$

When the atoms are far apart, the $E_{n,\underline{0}}$ are the atomic levels and the $u_{n,\underline{0}}(\underline{r})$ the atomic eigenfunctions. Nevertheless, this equation exhibits the symmetry of the crystal potential.

The $\underline{k} \cdot \underline{p}$ method assumes that the values $E_{n,\underline{0}}$ are known either from theory or experiment. In order to calculate the energies for $k \neq 0$, the $\underline{k} \cdot \underline{p}$ term is treated as a perturbation in the Hamiltonian, i.e., the method is only valid for small values of $\underline{k}$ close to $\underline{k} = 0$. The $\underline{k} \cdot \underline{p}$ term in the Hamiltonian commutes with the translational operator so that we can expand the eigenfunctions for finite crystal wavevectors as

$$u_{n,\underline{k}}(\underline{r}) \;=\; \sum_l c_l(\underline{k})\, u_{l,\underline{0}}(\underline{r})\,. \tag{4.18}$$

By inserting this expansion into Eq. 3.19, we obtain

$$\sum_l \left[\left(E_{l,\underline{0}} - E_{n,\underline{k}} + \frac{\hbar^2\, \underline{k}^2}{2\,m}\right) \delta_{nl} \;+\; \frac{\hbar\, \underline{k}}{m} \cdot \underline{p}_{nl}\right] c_l(\underline{k}) = 0\,, \tag{4.19}$$

where

$$\underline{p}_{nl} \;=\; \langle n\underline{0}|\underline{p}|l\underline{0}\rangle \;=\; \int d^3r\, u^*_{n,\underline{0}}\, \underline{p}\, u_{l,\underline{0}}\,. \tag{4.20}$$

Note that $\underline{p}_{nn} = 0$ due to the definite parity of the wave function. Assuming that the nth band edge is non degenerate, the expansion coefficients will have the following property

$$c_n(\underline{k}) \sim 1 \quad \text{and} \quad c_l(\underline{k}) \;=\; \underline{\alpha}_l \cdot \underline{k} \quad \text{for } l \neq n\,, \tag{4.21}$$

since $c_l(\underline{0}) = \delta_{nl}$. Using this knowledge in Eq. 4.19, we can determine the coefficient $\underline{\alpha}_l$ $(l \neq n)$

$$\underline{\alpha}_l = \frac{\hbar}{m} \frac{\underline{p}_{nl}}{E_{n,\underline{0}} - E_{l,\underline{0}}} . \tag{4.22}$$

Inserting this back into Eq. 4.19 leads to the energies up to second order in the crystal wavevector

$$E_{n,\underline{k}} = E_{n,\underline{0}} + \frac{\hbar^2 \underline{k}^2}{2m} + \frac{\hbar^2}{m^2} \sum_{l \neq n} \frac{|\underline{p}_{nl} \cdot \underline{k}|^2}{E_{n,\underline{0}} - E_{l,\underline{0}}} . \tag{4.23}$$

The effective masses at the band minima or maxima are readily determined from Eq. 4.23

$$\left(\frac{1}{m_n^*}\right)_{ij} = \frac{1}{m} \delta_{ij} + \frac{2}{m^2} \sum_{l \neq n} \frac{p_{nl}^i \, p_{nl}^j}{E_{n,\underline{0}} - E_{l,\underline{0}}} . \tag{4.24}$$

This method therefore allows the determination of the effective masses directly from the energy spectrum at $\underline{k} = 0$ and the matrix elements p_{nl}^i. To estimate the effective mass in a typical semiconductor, we can use a simplified version of Eq. 4.24

$$m^* = \frac{m E_G}{E_G + 2\langle p^2 \rangle / m} , \tag{4.25}$$

where E_G denotes the energy gap, which is typically of the order of 1 eV. $\langle p^2 \rangle / m$ can be approximated by $h^2/(ma^2)$, which is of the order of 12 eV. Therefore, a typical value for the effective mass is $0.04m$. Note that the effective mass increases with increasing energy gap. Looking only at two bands in a cubic crystal, the effective mass at the bottom of the conduction band and top of the valence band are given by

$$m_C^* = \frac{m E_G}{E_G + |\langle 1 | p_x | 2 \rangle|^2 / m} \quad \text{and}$$

$$m_V^* = \frac{m E_G}{E_G - |\langle 1 | p_x | 2 \rangle|^2 / m} . \tag{4.26}$$

Therefore, the sum of the inverse of the two effective masses is equal to $2/m$. The effective mass of the electron and the effective mass of the so-called light hole are quite similar in cubic crystals.

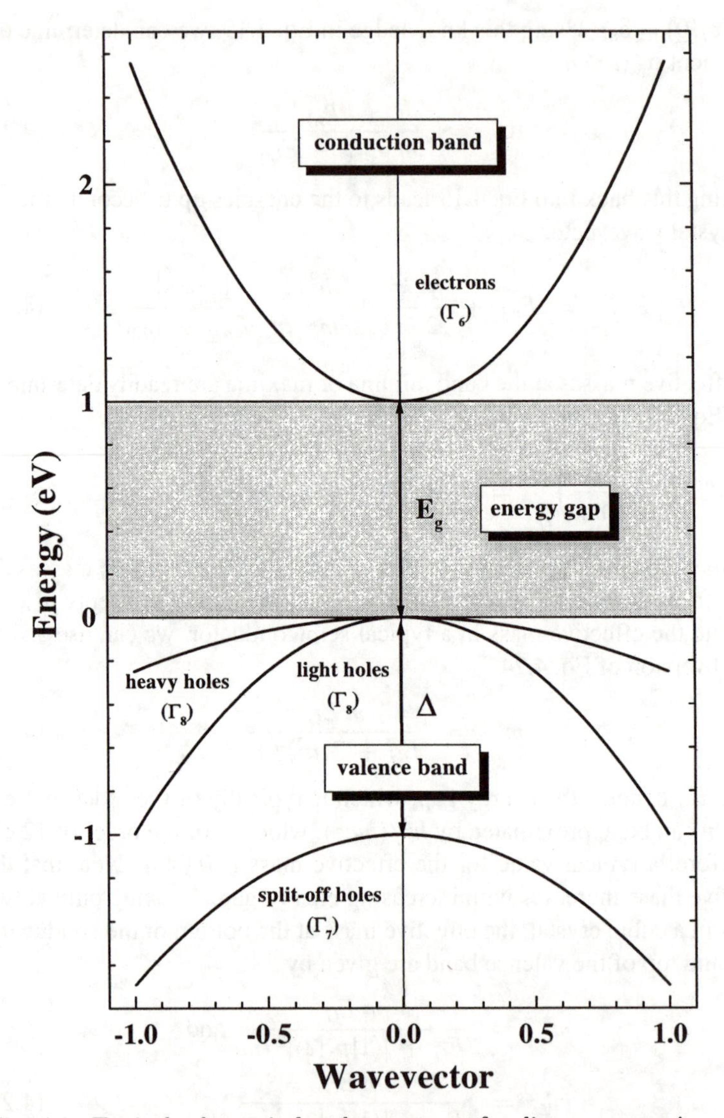

Fig. 4.1. Typical schematic band structure of a direct gap semiconductor with zincblende crystal structure in the vicinity of the Γ-point.

The valence band structure is more complicated due to the degeneracy of the top valence band states. In order to determine the full band structure near the energy gap for small crystal wavevectors, the angular momentum of the states has to be taken into account. In group IV and compound semiconductors, the upper valence band is formed by bonding p-like states, while the conduction band is formed by antibonding s-like states. The degeneracy of the top of the valence band is therefore three corresponding to the three p-orbitals, while the bottom of the conduction band has no degeneracy. As for isolated atoms, the coupling between the orbital angular momentum of these states and the electronic spin results in a splitting of the top valence band into heavy, light, and split-off holes. In cubic semiconductors, the heavy- and light-hole states remain degenerate at $\underline{k} = 0$, while the split-off holes are shifted to lower energies by an amount Δ, which is typically of the order of 0.1 to 1 eV. In Fig. 4.1 a typical schematic band structure near the Γ-point of a direct gap semiconductor with zincblende crystal structure is shown. It consists of a single conduction band and three valence bands labeled heavy holes, light holes and split-off holes. The effective masses of electrons and light holes are similar, while the heavy hole effective mass is much larger. Note that this band structure is only valid near the center of the Brillouin zone, i.e., for small crystal wavevectors. A model, where the effective masses of all four bands are expressed only through three parameters, P, E_G, and Δ, was developed by E.O. Kane. The following expressions for the energy dispersion can be used near $\underline{k} = 0$

$$
\begin{aligned}
\text{electrons:} \quad E(k) &= E_G + \frac{\hbar^2 k^2}{2m} + \frac{P^2 k^2}{3}\left(\frac{2}{E_G} + \frac{1}{E_G + \Delta}\right), \\
\text{heavy holes:} \quad E(k) &= -\frac{\hbar^2 k^2}{2m}, \\
\text{light holes:} \quad E(k) &= -\frac{\hbar^2 k^2}{2m} - \frac{2P^2 k^2}{3E_G}, \\
\text{split-off holes:} \quad E(k) &= -\Delta - \frac{\hbar^2 k^2}{2m} - \frac{P^2 k^2}{3(E_G + \Delta)} .
\end{aligned}
\tag{4.27}
$$

According to this model, the heavy hole effective mass is equal to the free electron mass, while electrons, light holes, and split-off holes have comparable masses

$$\begin{aligned}
\text{electrons:} \quad m_e^* &= \frac{m}{1+\frac{2\,m\,P^2}{3\,\hbar^2}\left(\frac{2}{E_G}+\frac{1}{E_G+\Delta}\right)}\,, \\
\text{heavy holes:} \quad m_{hh}^* &= m\,, \\
\text{light holes:} \quad m_{lh}^* &= \frac{m}{1+\frac{4\,m\,P^2}{3\,\hbar^2 E_G}}\,, \\
\text{split-off holes:} \quad m_{so}^* &= \frac{m}{1+\frac{2\,m\,P^2}{3\,\hbar^2 (E_G+\Delta)}}\,.
\end{aligned} \tag{4.28}$$

The parameter $P^2/\,\hbar^2$, which has the units of a velocity, can be expressed in terms of the energy gap, the effective electron mass, and the split-off hole energy Δ

$$\frac{P^2}{\hbar^2} = \frac{1}{2\,m}\left(\frac{m}{m_e^*}-1\right)\frac{E_G+\Delta}{1+\frac{2\,\Delta}{3\,E_G}}\,. \tag{4.29}$$

This parameter will be used in Chapter 10 to determine the value of the optical absorption coefficient in GaAs.

4. The Band Structure of Selected Semiconductors

In this section, the actual band structures of some selected semiconductors are discussed. The effective masses for many semiconductors listed in Chapter 1 are compiled. Note that some of these values change as time progresses. Group IV semiconductors exhibit an indirect energy gap, i.e., the maximum energy of the valence band and the minimum energy of the conduction band occur at different wavevectors, i.e., different locations of the first Brillouin zone. For semiconductors, the maximum energy of the valence band is always at the center of the Brillouin zone implying that an indirect energy gap is connected with a minimum energy of the conduction band at $\underline{k} \neq 0$. In Fig. 4.2, a typical band structure for an indirect semiconductor such as Si or Ge is shown. In Si, the minimum of the conduction band occurs near the X-point (about $0.8\times 2\pi/a$), while in Ge the minimum is at the L-point. The band structure near the absolute minimum of the conduction band is not isotropic so that two different effective masses appear,

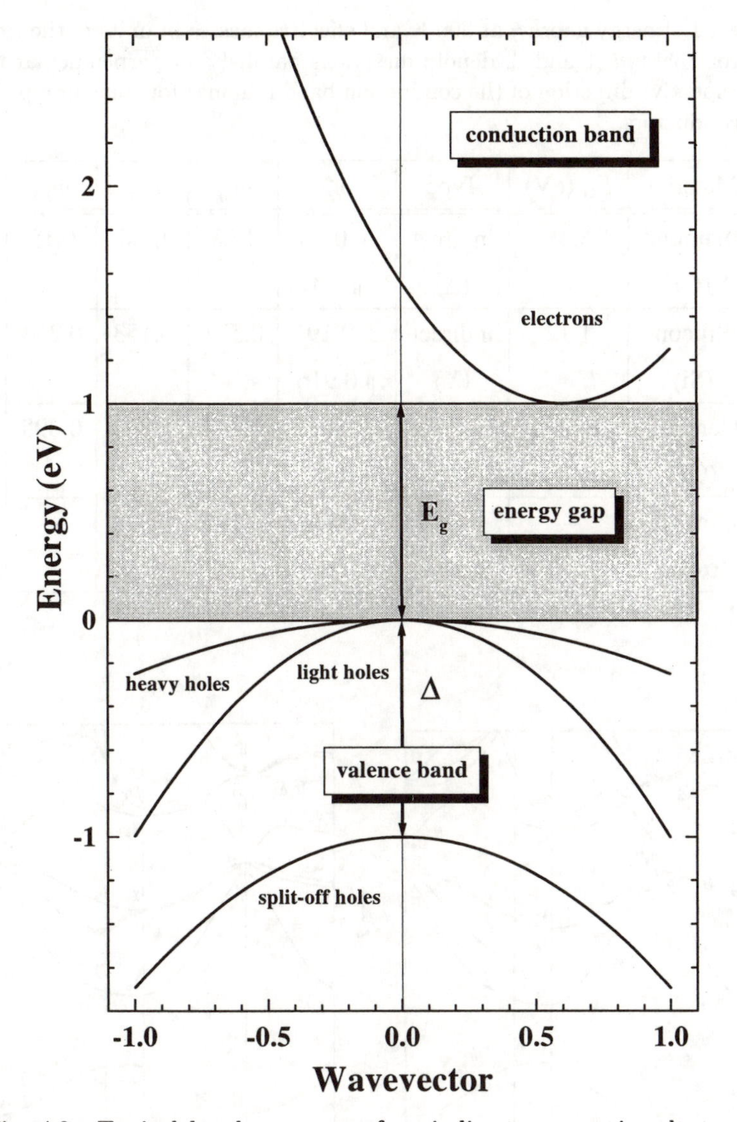

Fig. 4.2. Typical band structure of an indirect gap semiconductor with diamond crystal structure.

Table 4.1. Energy gap E_G at 300 K and effective masses in units of the free electron mass m ($\parallel$ and $\perp$ denote directions parallel and perpendicular to the respective direction of the conduction band minima) for some group IV semiconductors.

Material	E_G (eV)	Type	m_e^*	m_{hh}^*	m_{lh}^*	m_{so}^*
Diamond (C)	5.48	indirect (Δ)	$\perp$ 0.36 $\parallel$ 1.4	1.08	0.36	0.15
Silicon (Si)	1.12	indirect (X)	$\perp$ 0.191 $\parallel$ 0.916	0.537	0.153	0.234
German. (Ge)	0.664	indirect (L)	$\perp$ 0.081 $\parallel$ 1.59	0.284	0.044	0.095
Gray tin (α-Sn)	—	—	0.0236	0.195	0.058	—

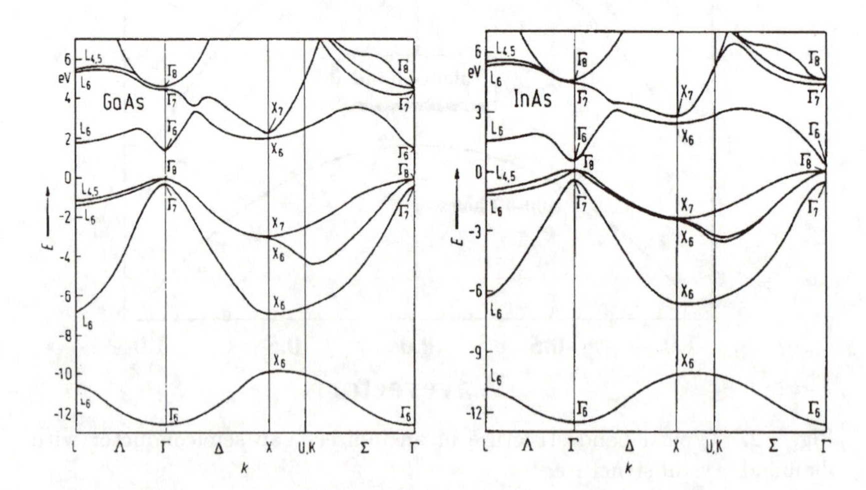

Fig. 4.3. Band structure of GaAs (left) and InSb (right) taken from *Landolt-Börnstein, Series III, Vol. 22a.*

parallel and perpendicular to the directions of the respective conduction band minima (e.g., in Si the $\langle 100 \rangle$ directions). Therefore, two effective electron masses are listed in Tab. 4.1 for the two group IV semiconductors. The band structure of α-Sn is very different from the typical group IV semiconductor. The s-like conduction band edge is energetically below the p-like valence band maximum. This causes an inversion in the curvature

Table 4.2. Energy gap E_G at 300 K and effective masses in units of the free electron mass m ($\parallel$ and $\perp$ denote directions parallel and perpendicular to the respective direction of the conduction band minima) for III-V semiconductors.

Material	E_G (eV)	Type	m_e^*	m_{hh}^*	m_{lh}^*	m_{so}^*
BN	6.4	indirect	0.752	0.38	0.15	—
GaN	3.44	direct	0.20	0.80	—	—
AlP	2.51	indirect (X)	$\perp$ 0.212 $\parallel$ 3.67	0.51	0.21	0.29
GaP	2.27	indirect (Δ)	$\perp$ 0.25 $\parallel$ 7.25	0.67	0.17	0.46
AlAs	2.15	indirect (X)	$\perp$ 0.19 $\parallel$ 1.1	0.41	0.15	0.24
InN	1.89	direct	0.12	0.50	0.17	—
AlSb	1.62	indirect (Δ)	$\perp$ 0.26 $\parallel$ 1.8	0.34	0.12	0.29
GaAs	1.424	direct	0.063	0.50	0.076	0.145
InP	1.34	direct	0.079	0.60	0.12	0.12
GaSb	0.75	direct	0.041	0.28	0.05	0.15
InAs	0.354	direct	0.024	0.41	0.024	0.14
InSb	0.18	direct	0.014	0.42	0.016	0.43

of the light hole band. Therefore, α-Sn is a zero-gap semiconductor with its lowest conduction and highest valence band being degenerated at Γ.

Most III-V semiconductors with a direct energy gap, which crystallize in the zincblende structure, exhibit the schematic band structure shown in Fig. 4.1. In order to demonstrate the band structure further away from the energy gap, two typical band structures are shown in Fig. 4.3, GaAs on the left and InSb on the right. Except for the different energy scales, the overall band structures are very similar. Only the value of the energy gap and the effective masses differ significantly as given in Tab. 4.2. Note that cubic III-V semiconductors with large energy gaps exhibit an indirect band structure, while for GaAs and all other III-V compounds with lower energy gap the band structure becomes direct. For the direct-gap semiconductors, the effective electron mass decreases with decreasing gap energy.

Table 4.3. Energy gap E_G at 300 K (* denotes value at low temperature) and effective masses in units of the free electron mass m for II-VI semiconductors.

Material	E_G (eV)	Type	m_e^*	m_h^*
ZnS (c)	3.68	direct	0.34	1.76
ZnS (h)	3.91	direct	0.28	0.49
ZnO	3.44*	direct	0.28	0.59
ZnSe (c)	2.7	direct	0.16	0.78, 0.145
CdS (c)	2.55	direct	0.14	0.51
CdS (h)	2.51	direct	0.21	0.68
ZnTe	2.28	direct	0.12	0.6
CdSe (c)	1.9*	direct	0.11	0.44
CdSe (h)	1.75	direct	0.11	0.45
CdTe	1.475	direct	0.096	0.63
HgSe	—	—	—	0.78
HgTe	—	—	0.031	0.32

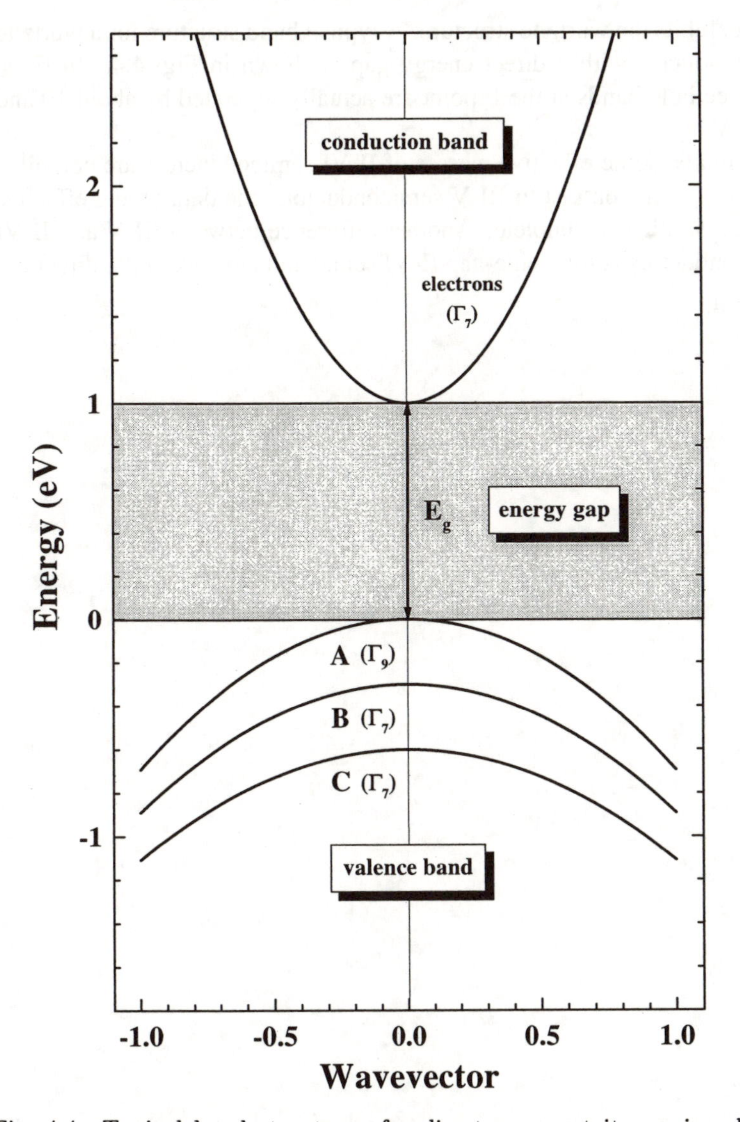

Fig. 4.4. Typical band structure of a direct-gap wurtzite semiconductor such as hexagonal GaN.

GaN exhibits the wurtzite structure. A typical band structure for a wurtzite semiconductor with a direct energy gap is shown in Fig. 4.4. In GaN, the three hole bands at the Γ point are actually separated by about 10 and 35 meV.

Finally, some effective masses of II-VI semiconductors are compiled in Tab. 4.3. In contrast to III-V semiconductors, the data on the effective masses is rather incomplete. Another difference between III-V and II-VI semiconductors is that wide-gap II-VI semiconductors exhibit a direct energy gap.

CHAPTER 5

DENSITY OF STATES AND CARRIER STATISTICS

Electrons and holes are Fermions. This implies that two electrons (or holes) cannot occupy the same energy state. Neglecting spin degeneracy, each energy state can contain at most one electron. At zero temperature, the states will therefore fill up to a certain wavevector, which is called the Fermi wavevector $\underline{k}_F$. For electrons (holes) near the bottom (top) of the conduction (valence) band, the Fermi wavevector is associated with a Fermi energy E_F defined through

$$E_F \quad = \quad \frac{\hbar^2 \, \underline{k}_F^2}{2 \, m^*} \, . \tag{5.1}$$

A very important concept for determining the carrier density, the total energy, and the absorption coefficient in semiconductors is the density of states. Since the states are occupied up to a certain wavevector or energy, it is essential to determine, how the states are distributed in energy.

1. The Density of States

The density of states is simply obtained by counting the number of possible states within a certain energy band

$$G_n(E) \quad = \quad \sum_{\underline{k}} \delta[E \; - \; E_n(\underline{k})] \, , \tag{5.2}$$

where $\delta(x - x_0)$ denotes the delta function, which is zero except for $x = x_0$. $G_n(E)$ is known as the density of states of the nth energy band and $E_n(\underline{k})$ as the energy of the respective state in the nth energy band. Since electrons are Fermions, i.e., they are spin $1/2$ particles, there are two possible states for a given energy, spin up and spin down. Therefore, the degeneracy for electrons and holes is 2. This can be generalized by introducing a degeneracy factor g in front of the sum in Eq. 5.2

$$G_n(E) = g \sum_{\underline{k}} \delta[E - E_n(\underline{k})] \,. \tag{5.3}$$

The total density of states is obtained by summing over all energy bands

$$G(E) = \sum_n G_n(E) \,. \tag{5.4}$$

In order to calculate the density of states per unit volume $g(E) = G(E)/V$ near the bottom of the conduction band (top of the valence band), we will take the continuum limit of Eq. 5.3 (in three dimensions)

$$g(E) = 2 \int \frac{d^3k}{(2\pi)^3} \, \delta[E - E(\underline{k})] \,, \tag{5.5}$$

where V denotes the volume of the crystal. This limit is justified, since there is a very large number of states in an energy band due to the large number of atoms in a macroscopic solid. Furthermore, we assume the simple parabolic dispersion near the bottom of the conduction band (top of the valence band) as outlined in the previous chapter. For a single band with an isotropic effective mass, the energy is given by

$$E_n(\underline{k}) = E_n(\underline{0}) + \frac{\hbar^2 \, \underline{k}^2}{2\, m^*} \,. \tag{5.6}$$

In the case of an isotropic mass, the volume element of k-space in Eq. 5.5 can be replaced by $d^3k = 4\pi k^2 dk$. For the integration over dk, we have to recall two properties of the δ-function

$$\begin{aligned} \delta(ax) &= \frac{1}{|a|}\, \delta(x) \quad \text{and} \\ \delta(x^2 - b^2) &= \frac{1}{2|b|}\, [\delta(x-b) + \delta(x+b)] \,. \end{aligned} \tag{5.7}$$

Since for the integration over dk the limits of the integral are 0 and ∞, only the first term of the sum in the second equation of Eq. 5.7 contributes.

Putting everything together, one obtains the following result for the density of states of electrons near the conduction band minimum in three dimensions

$$g_n(E) \;=\; \frac{(2m^*)^{3/2}}{2\pi^2\,\hbar^3}\,\sqrt{E - E_n}\,\Theta(E - E_n)\,, \tag{5.8}$$

where $\Theta(E - E_n)$ denotes the unit-step function, i.e.,

$$\Theta(E - E_n) \;=\; \begin{cases} 0 & \text{for} \quad E < E_n \\ 1 & \text{for} \quad E > E_n \end{cases} . \tag{5.9}$$

This density of states is valid for electrons near the conduction band minimum. In order to use this result for holes near the valence band maximum, one has to replace in Eq. 5.8 $E - E_n$ by $E_n - E$. If the band structure is not isotropic, $(m^*)^{3/2}$ has to be replaced by $\sqrt{m_x^*\, m_y^*\, m_z^*}$.

In one or two dimensions, the density of states can be obtained in a similar way as for three dimensions. However, the reduced dimensionality of k-space has to be taken into account in the integral

$$g(E) \;=\; 2\int \frac{d^n k}{(2\pi)^n}\,\delta[E - E(\underline{k})]\,, \tag{5.10}$$

where $n = 1$ or 2 depending on the dimension of the considered system. Performing the same calculation as in three dimensions, one obtains in two dimensions a constant density of states , i.e.,

$$g_n(E) \;=\; \frac{m^*}{\pi\,\hbar^2}\,\Theta(E - E_n)\,. \tag{5.11}$$

For a non-isotropic effective mass, m^* has to be replaced by $\sqrt{m_x^*\, m_y^*}$. This density of states can be directly observed for the 2-dimensional electron gas in field effect transistors or quantum wells. In one dimension, the density of states diverges for E approaching E_n according to

$$g_n(E) \;=\; \frac{\sqrt{2m^*}}{\pi\,\hbar}\,\frac{\Theta(E - E_n)}{\sqrt{E - E_n}}\,. \tag{5.12}$$

A one-dimensional system can be realized by a semiconductor quantum wire or by a bulk semiconductor experiencing a magnetic field. For a

Table 5.1. Density of states for electrons near the conduction band minimum in semiconductors with different dimensions.

Dimension	Prefactor	Energy dependence
3	$\frac{(2m^*)^{3/2}}{2\,\pi^2\,\hbar^3}$	$\sqrt{E-E_n}\,\Theta(E-E_n)$
2	$\frac{m^*}{\pi\,\hbar^2}$	$\Theta(E-E_n)$
1	$\frac{\sqrt{2m^*}}{\pi\,\hbar}$	$\frac{\Theta(E-E_n)}{\sqrt{E-E_n}}$
0	2	$\delta(E-E_n)$

zero-dimensional system, e.g., a semiconductor quantum dot, the density of states can be directly derived from its definition in Eq. 5.2

$$g_n(E) \;=\; 2\,\delta(E - E_n)\,. \tag{5.13}$$

The density of states for isotropic systems of different dimensions are listed in Tab. 5.1 in terms of the prefactor and their energy dependence. The respective density of states is shown versus energy for an isotropic effective mass in Fig. 5.1. Tab. 5.2 lists the density of states in two and three dimensions for electrons near the conduction band minimum in the case of

Table 5.2. Density of states for electrons near the conduction band minimum in non-isotropic semiconductors in 2 and 3 dimensions.

Dimension	Prefactor	Energy dependence
3	$\frac{(8m_x^*m_y^*m_z^*)^{1/2}}{2\,\pi^2\,\hbar^3}$	$\sqrt{E-E_n}\,\Theta(E-E_n)$
2	$\frac{(m_x^*m_y^*)^{1/2}}{\pi\,\hbar^2}$	$\Theta(E-E_n)$

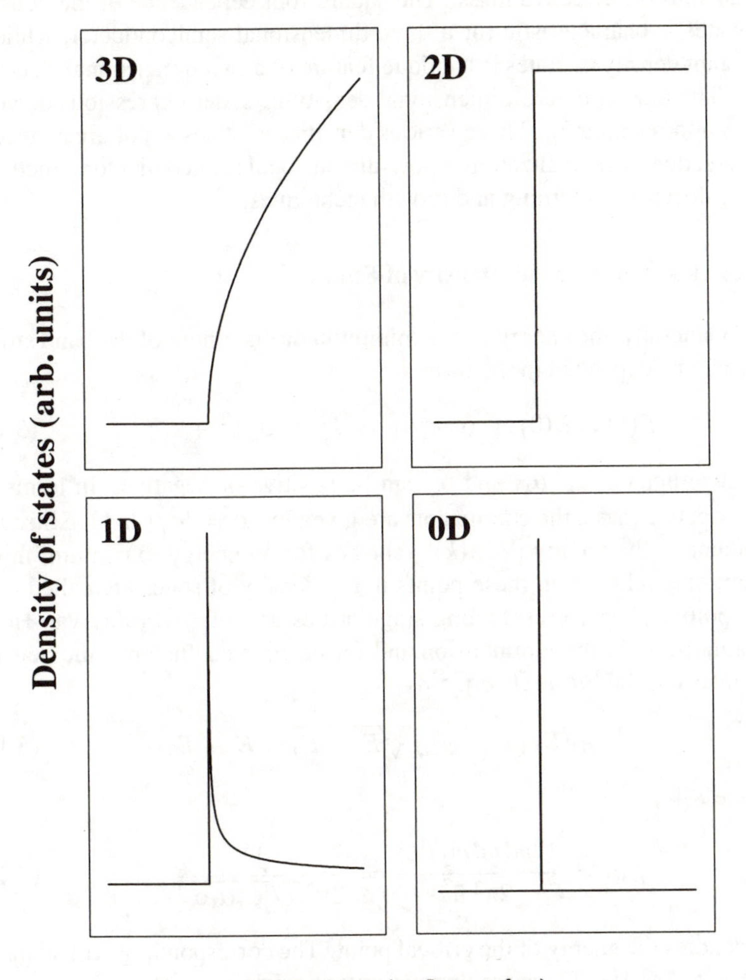

Fig. 5.1. Density of states for electrons in three-, two-, one-, and zero-dimensional semiconductors.

an anisotropic effective mass. The square-root dependence of the density of states is characteristic for a three-dimensional semiconductor, while a constant density of states is a unique feature of a two-dimensional electron gas. The one- and zero-dimensional density-of-states expressions diverge for E approaching E_n. These various densities of states are of great current interest due to the realization of low-dimensional semiconductor structures using different patterning and growth techniques.

2. Critical Points in the Density of States

Generally, the energy near a minimum or maximum of the band structure can be expanded in the form

$$E(\underline{k}) = E(0) + \alpha_x k_x^2 + \alpha_y k_y^2 + \alpha_z k_z^2 + \dots . \tag{5.14}$$

The coefficients α_x, α_y, and α_z can be positive or negative. In terms of the effective mass, the coefficients are given by $\alpha_i = \hbar^2/(2m_i^*)$. Since the gradient of the energy $|\nabla_{\underline{k}} E(\underline{k})|$ vanishes for an energy maximum, minimum, or saddle point, these points in the density of states are called critical points. The corresponding singularities are referred to as Van-Hove singularities. In three dimension and for positive coefficients, the density of states has the form (cf. Eq. 5.8)

$$g_{3D}(E) \quad = \quad g_{3D} \sqrt{E - E_0}\, \Theta(E - E_0) \,, \tag{5.15}$$

where

$$g_{3D} = \frac{(8m_x^* m_y^* m_z^*)^{1/2}}{2\pi^2 \hbar^3} = \frac{1}{2\pi^2 \sqrt{|\alpha_x \alpha_y \alpha_z|}} \,. \tag{5.16}$$

E_0 denotes the energy of the critical point. The corresponding critical point is denoted as M_0. There are three more critical points in three dimensions, M_1, M_2, and M_3. The subscript indicates the number of negative coefficients α_i. For M_1, α_x is negative, while α_y and α_z are positive. In the case of M_2, α_x and α_y are negative, while α_z is positive. For M_3, all three coefficients are negative. In two dimensions there are three critical points. The density of states for the M_0 critical point is given by (cf. Eq. 5.11)

$$g_{2D}(E) \quad = \quad g_{2D}\, \Theta(E - E_0) \tag{5.17}$$

Table 5.3. Critical points in the density of states for three, two, and one dimension. The corresponding energy dependence is also listed.

Dimensions	Type	Prefactor	Energy dependence
3	M_0	g_{3D}	$\sqrt{E-E_0}\,\Theta(E-E_0)$
	M_1	g_{3D}	$C-\sqrt{E_0-E}\,\Theta(E_0-E)$
	M_2	g_{3D}	$C-\sqrt{E-E_0}\,\Theta(E-E_0)$
	M_3	g_{3D}	$\sqrt{E_0-E}\,\Theta(E_0-E)$
2	M_0	g_{2D}	$\Theta(E-E_0)$
	M_1	$\frac{g_{2D}}{\pi}$	$-\ln\lvert E-E_0\rvert$
	M_2	g_{2D}	$\Theta(E_0-E)$
1	M_0	g_{1D}	$\frac{\Theta(E-E_0)}{\sqrt{E-E_0}}$
	M_1	g_{1D}	$\frac{\Theta(E_0-E)}{\sqrt{E_0-E}}$

with $g_{2D} = (2\pi\,\sqrt{|\alpha_x\alpha_y|})^{-1}$. In one dimension, there are two critical points. The density of states for the M_0 critical point is given by (cf. Eq. 5.12)

$$g_{1D}(E) \quad = \quad g_{1D}\,\frac{\Theta(E\;-\;E_0)}{\sqrt{E\;-\;E_0}} \tag{5.18}$$

with $g_{1D} = (\pi\,\sqrt{|\alpha_x|})^{-1}$. The critical points in the density of states are listed in Tab. 5.3 for the different dimensions. In Fig. 5.2, their respective energy dependence is shown for three, two, and one dimension.

3. The Fermi-Dirac Distribution

Electrons are Fermions. At zero temperature, all states below the Fermi energy are occupied, and all states above the Fermi energy are empty. Therefore, the distribution function, i.e., the probability of occupancy of an energy state, can be written at zero temperature as

$$f_{FD}(E,0) \quad = \quad \Theta(E_F\;-\;E)\,. \tag{5.19}$$

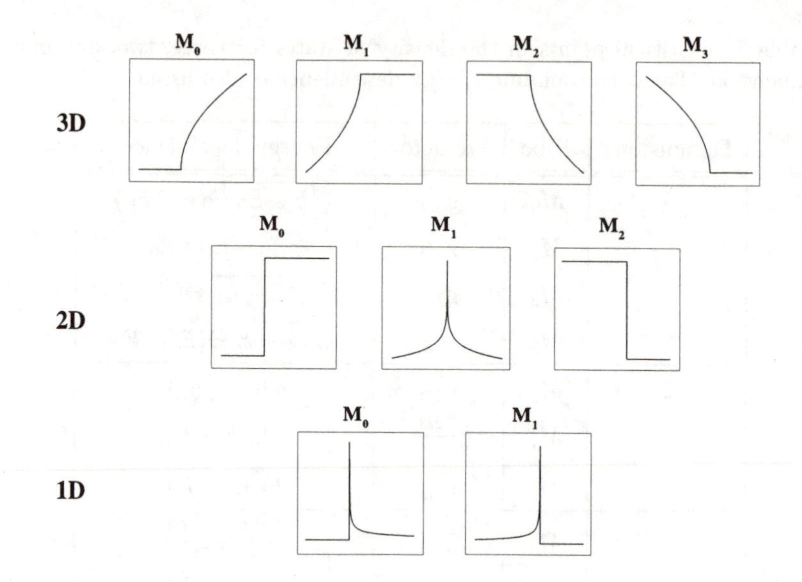

Fig. 5.2. Critical points of the density of states in three, two, and one dimension.

At finite temperature T, this unit step function smears out, and the electron distribution follows the Fermi-Dirac distribution f_{FD}

$$f_{FD}(E,T) = \frac{1}{1 + \exp[(E-\mu)/k_B T]}, \tag{5.20}$$

where μ denotes the chemical potential and k_B the Boltzmann constant. When T approaches zero, the distribution in Eq. 5.19 is recovered. With increasing temperature, the transition region widens as shown in Fig. 5.3, where f_{FD} is plotted versus energy for several temperatures in units of the energy scale on the bottom.

The width of the Fermi-Dirac distribution as defined by $0.12 \leq f_{FD} \leq 0.88$ corresponds to $4k_BT$. At 3 K, it has a value of about 1 meV, while at room temperature (300 K) it is 103 meV. At $T = 0$, the chemical potential is equal to the Fermi energy. However, at finite temperature, these two quantities are not exactly the same. For most applications in semiconductors, the term chemical potential is replaced by Fermi energy, although this

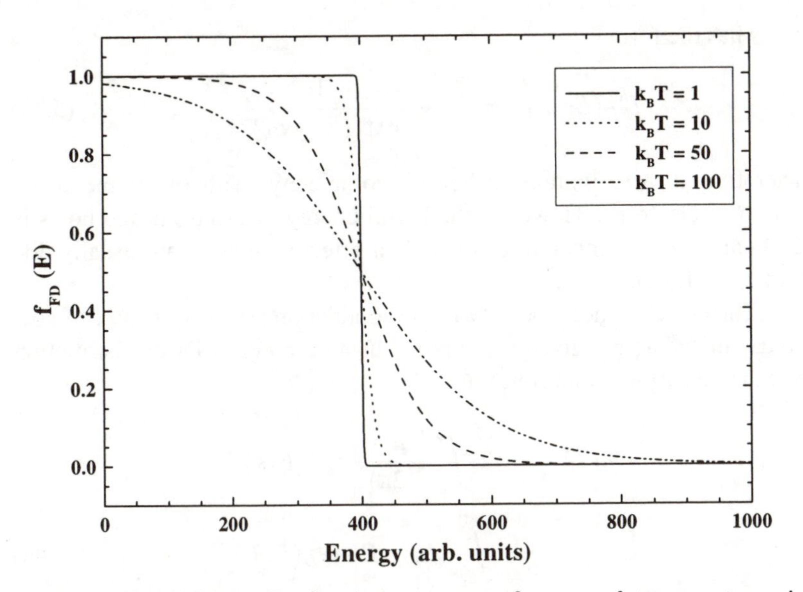

Fig. 5.3. Fermi-Dirac distribution vs energy for several temperatures in units of the energy scale at the bottom of the graph.

is not strictly correct. In the limit that the relevant energies are much larger than the Fermi energy, i.e., ($E - E_F \gg k_B T$), the Fermi-Dirac distribution can be approximated by the Maxwell-Boltzmann distribution f_{MB}

$$f_{MB}(E,T) \quad = \quad \exp(\mu/k_BT)\exp(-E/k_BT)\,. \tag{5.21}$$

In this limit, $f_{FD} \ll 1$, and the occupancy of an energy state is very small. In other terms, the exclusion principle is not effective anymore, since the probability of occupying a state with two electrons is negligible. From a statistical point of view, the electron behaves like a classical particle.

The occupation of hole states f_{FD}^h can be derived from the occupation of electrons states f_{FD}^e by

$$\begin{aligned} f_{FD}^h(E,T) &= 1 - f_{FD}^e(E,T) \\ &= \frac{1}{\exp[(\mu - E)/k_BT] + 1}\,. \end{aligned} \tag{5.22}$$

Since the energy for holes is measured from the top of the valence band into the valence band, i.e., $E_h = -E_e$ and $\mu_h = -\mu_e$, the distribution function for holes has the form

$$f_{FD}^{h}(E_h,T) = \frac{1}{1 + \exp[(E_h - \mu_h)/k_B T]} . \tag{5.23}$$

Therefore, the distribution for holes is completely analogous to the distribution of electrons. However, the Fermi energy of electrons and holes is different at zero temperature, since their effective masses are usually different (cf. Eq. 5.1).

The electron (hole) density in the conduction (valence) band can now be calculated at any given temperature using the Fermi-Dirac distribution and the density-of-state concept.

$$\begin{aligned} n(T) &= 2 \int_0^\infty \frac{d^3k}{(2\pi)^3} f_{FD}(E(\underline{k}),T) \\ &= \int_{E_C}^\infty dE\, g(E)\, f_{FD}(E,T) . \end{aligned} \tag{5.24}$$

Solving this integral for an isotropic effective mass leads to

$$n(T) = N_C(T)\, F_{1/2}[(\mu - E_C)/k_B T] , \tag{5.25}$$

where the effective density of states $N_C(T)$ is defined as

$$N_C(T) = 2 \left(\frac{m_e^*\, k_B T}{2\, \pi\, \hbar^2} \right)^{3/2} \tag{5.26}$$

and

$$F_{1/2}(y) = \frac{2}{\sqrt{\pi}} \int_0^\infty dx \frac{x^{1/2}}{\exp(x-y) + 1} . \tag{5.27}$$

For an effective mass of 1, the value of $N_C(T)$ at 300 K is 2.51×10^{19} cm^{-3} including the spin degeneracy factor. If the electrons (holes) can be described by Maxwell-Boltzmann statistics, the function in Eq. 5.27 is replaced by $\exp(y)$, and the carrier density is given by

$$n(T) = N_C(T) \exp[(\mu - E_C)/k_B T] . \tag{5.28}$$

A similar result is obtained for holes

$$p(T) \;=\; N_V(T)\exp[(E_V-\mu)/k_BT]\,. \tag{5.29}$$

At zero temperature, the conduction band is completely empty, and no holes occupy the valence band. However, at finite temperature, electrons (holes) can be thermally excited across the energy gap. Since in an intrinsic semiconductor the electron and holes density are equal at any temperature, the chemical potential or Fermi energy is given by

$$\mu(T) \;=\; \frac{1}{2}\,(E_C+E_V) + \frac{3}{4}\,k_BT\,\ln\left(\frac{m_h^*}{m_e^*}\right)\,. \tag{5.30}$$

For equal effective masses, the chemical potential becomes independent of temperature. At zero temperature, the chemical potential is equal to the Fermi energy, which according to Eq. 5.30 corresponds for intrinsic semiconductors to $(E_C+E_V)/2=(E_C-E_V)/2+E_V=E_G/2+E_V$. The Fermi level of an intrinsic semiconductor is exactly located between the top of the valence band and the bottom of the conduction band at zero temperature. Since usually the effective mass of holes is larger than that one of electrons, the chemical potential increases with increasing temperature. However, for a typical direct gap semiconductor such as GaAs, this correction only amounts to 34 meV at room temperature, which is still a factor of 20 smaller than $E_G/2=710$ meV.

4. Intrinsic Semiconductors

The intrinsic carrier concentration $n_i=p_i$ is obtained by multiplying Eq. 5.28 with Eq. 5.29

$$\begin{aligned} n_i(T)^2 &= n_i(T)\,p_i(T) \\ &= N_C(T)\,N_V(T)\exp[-(E_C-E_V)/k_BT]\,. \end{aligned} \tag{5.31}$$

Taking the square root of Eq. 5.31, the intrinsic electron and hole concentrations are given by

$$\begin{aligned} n_i(T) &= p_i(T) \\ &= \sqrt{N_C(T)\,N_V(T)}\exp[-E_G/(2k_BT)]\,. \end{aligned} \tag{5.32}$$

The temperature dependence of the intrinsic carrier concentration follows therefore $T^{3/2} \exp[-E_G/(2k_BT)]$. The dominant contribution to the temperature dependence originates from the exponential function. Plotting $\ln(n_i)$ versus $1/T$ results in an almost straight line with a slope corresponding to one half of the gap energy.

In the case of an indirect band gap semiconductor such as Si or Ge, the effective mass of the conduction band is not isotropic. Furthermore, in Si there are six degenerate minima near the X-point, while in Ge the degeneracy is four. Therefore, the isotropic effective mass $(m^*)^{3/2}$ in the density of states has to be replaced by $\sqrt{(m_\perp^*)^2\, m_\parallel^*}$. In addition, the degeneracy factor is changed. For Si, the effective density of states $N_C(T)$ near the conduction band edge is given by

$$N_C(T) \;=\; 12\,\sqrt{(m_\perp^*)^2\, m_\parallel^*}\left(\frac{k_BT}{2\,\pi\,\hbar^2}\right)^{3/2} \;. \tag{5.33}$$

For the valence band, heavy- and light-hole states are usually degenerate at $\underline{k}=\underline{0}$. The contribution from heavy- and light-hole states have to be added, i.e., the effective mass $(m_h^*)^{3/2}$ has to be replaced by $(m_{hh}^*)^{3/2} + (m_{lh}^*)^{3/2}$. The effective density of states for the conduction and valence band as well as the intrinsic carrier concentration in Ge, Si, and GaAs at 300 K are listed in Tab. 5.4. The temperature dependence of $n_i(T)$ in Ge, Si, and GaAs is shown in Fig. 5.4, where $\ln(n_i)$ is plotted versus $1/T$. Ge exhibits the largest intrinsic carrier density of these three materials as a consequence of the smallest energy gap.

Table 5.4. Band edge effective density of states for the conduction N_C and valence N_V band, gap energy E_G, and intrinsic carrier concentration n_i in Ge, Si, and GaAs at 300 K.

Material	N_C (cm^{-3})	N_V (cm^{-3})	E_G (eV)	n_i (cm^{-3})
Ge	1.03×10^{19}	4.03×10^{18}	0.664	1.71×10^{13}
Si	2.75×10^{19}	1.14×10^{19}	1.12	6.93×10^{9}
GaAs	3.97×10^{17}	9.39×10^{18}	1.424	2.11×10^{6}

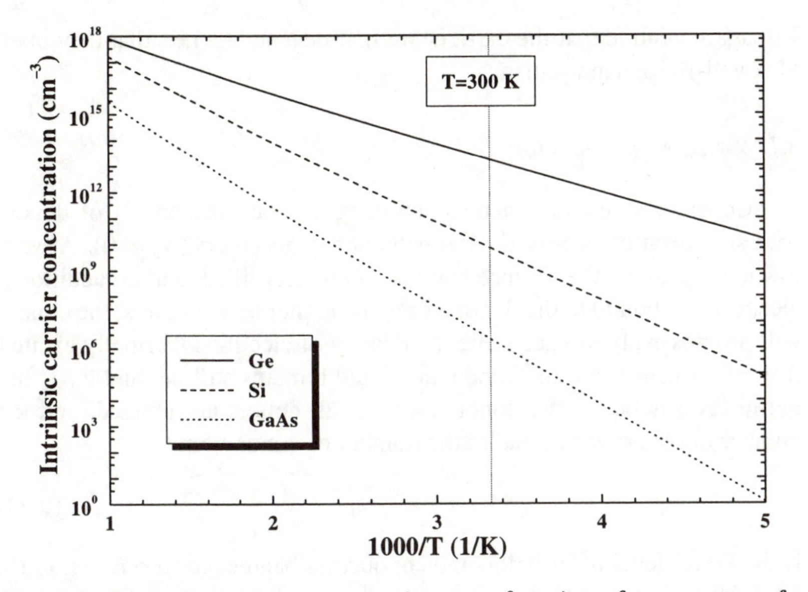

Fig. 5.4. Intrinsic carrier concentration as a function of temperature for Ge, Si, and GaAs.

5. Extrinsic Semiconductors

The intrinsic carrier densities at room temperature listed in Tab. 5.4 are much too small to result in a significant conductivity. The real difference between semiconductors and metals is that one can dope semiconductors at almost any level resulting in a very large tunability of the carrier density over many orders of magnitude. If some atoms in a Si crystal are replaced by group III atoms, there will be a shortage of electrons resulting in free holes in the valence band. Therefore, aluminum acts as an acceptor in Si, while phosphorus supplies one extra electron acting as a donor. If only 10^{-6} of all Si atoms are replaced by donor or acceptor atoms, the doping and resulting carrier density will be of the order of 10^{17} cm^{-3}, since each donor (acceptor) atom supplies one electron (hole). This carrier density is already much larger than the intrinsic carrier density. However, we again have to consider the carrier statistics in order to determine the actual electron (hole) density in the conduction (valence) band. In the following, we will again assume that the carriers are non-degenerate, i.e., that they obey Maxwell-Boltzmann statistics.

5.1. N-type semiconductors

Consider a semiconductor containing a concentration N_d of donors with an ionization energy E_d, but without any acceptors ($N_a = 0$). At very low temperatures, the valence band is completely filled, and all additional electrons are bound to the donor atoms. At higher temperatures, the donors will progressively ionize, while the density of electrons thermally excited from the valence into the conduction band remains still negligible. If the Fermi level is below the donor level E_d, all donors are ionized, and the number of electrons is equal to the number of donor atoms

$$n = N_d . \tag{5.34}$$

If the Fermi level is far below the conduction band edge ($\mu - E_C \ll k_B T$), Eq. 5.28 can be used to determine the chemical potential

$$\mu - E_C = k_B T \ln\left(\frac{N_d}{N_C}\right) . \tag{5.35}$$

It follows that for a doping density of 10^{16} cm^{-3} the Fermi level (or chemical potential) is about 180 meV below the conduction band edge at room temperature. For increasing temperature, the Fermi level moves further away from the conduction band edge. However, at very high temperatures, the intrinsic contribution of electrons from the valence band p cannot be neglected anymore. The number of conduction electrons is then given by

$$n = N_d + p . \tag{5.36}$$

The relation $np = n_i^2$ is valid even for extrinsic semiconductors as long as we can use Maxwell-Boltzmann statistics, i.e., in the non-degenerate limit. Therefore, Eq. 5.36 can be rewritten in following way

$$n = \frac{1}{2} N_d + \sqrt{\frac{N_d^2}{4} + n_i^2} . \tag{5.37}$$

If the intrinsic carrier density is small, we can expand this expression to estimate the hole density

$$n \approx N_d + \frac{n_i^2}{N_d} . \tag{5.38}$$

Therefore, the hole density is approximately

$$p \approx \frac{n_i^2}{N_d} . \tag{5.39}$$

The temperature range, for which the electron density is approximately equal to the donor concentration, is called the saturation regime. In this regime, the number of holes is much smaller than the number of intrinsic electrons n_i. The ratio of the number of electrons to the number of holes

$$\frac{n}{p} = \frac{N_d^2}{n_i^2} \tag{5.40}$$

is very large in this case. In Si for a doping density of $N_d = 10^{16}$ cm^{-3}, this ratio is 10^{12} at 300 K. Therefore, the electrons are called majority carriers and the holes are minority carriers. At very high temperatures, when $n_i \gg N_d$, the intrinsic regime is recovered

$$\begin{aligned} n &\approx n_i + \frac{N_d}{2} \quad \text{and} \\ p &\approx n_i - \frac{N_d}{2} . \end{aligned} \tag{5.41}$$

5.2. P-type semiconductors

The same consideration as for a purely n-type semiconductor can also be applied to purely p-type semiconductors, which are only doped with acceptors. There is a symmetry between purely n-type and purely p-type semiconductors. In Tab. 5.5, the corresponding quantities and equations are listed. In p-type semiconductors, the holes are the majority carriers, and the electrons are the minority carriers. The equations listed in Tab. 5.5 can either be derived from hole statistics or by replacing n by p, N_d by N_a,

Table 5.5. Comparison between n- and p-type semiconductors .

Quantity	N-type (donors)	P-type (acceptors)
Fermi level	$\mu = E_C +$ $k_BT \ln(N_d/N_C)$	$\mu = E_V -$ $k_BT \ln(N_a/N_V)$
Room temperature	$n = N_d$	$p = N_a$
Higher temperature	$n = \frac{N_d}{2} + \sqrt{\frac{N_d^2}{4} + n_i^2}$	$p = \frac{N_a}{2} + \sqrt{\frac{N_a^2}{4} + n_i^2}$
Very high temperatures	$n \approx n_i + \frac{N_d}{2}$ $p \approx n_i - \frac{N_d}{2}$	$p \approx n_i + \frac{N_a}{2}$ $n \approx n_i - \frac{N_a}{2}$

E_C by E_V, and N_C by N_V. There is also a change in the sign in front of k_BT in the equation for the chemical potential. The conductivity is p-type extrinsic in this case. We will see later that for such a semiconductor the apparent current is produced by positive electrons, i.e., the holes.

5.3. Compensated semiconductors

Although crystals of very high purity can be produced, there are always impurities incorporated during crystal growth. Therefore, even in a highly n-doped crystal, there is a certain amount of acceptors present and vice versa. This implies that intrinsic semiconductors, i.e., not intentionally doped semiconductors, are not intrinsic in the sense discussed above, but that they are actually compensated. In a fully compensated material, the electron and hole concentration are equal (as in the intrinsic case). However, the electrons and holes originate from donors and acceptors in this case. Assuming a certain donor density N_d and acceptor density N_a with the condition $N_d > N_a$, the electron density in the saturation regime is obtained by taking the difference between the donor and acceptor density

$$n = N_d - N_a . \tag{5.42}$$

The electron density is reduced by the acceptor density, and the n-type semiconductor is partially compensated. The chemical potential decreases, since the electron density is reduced

$$\mu = E_C + k_B T \ln\left(\frac{N_d - N_a}{N_C}\right). \tag{5.43}$$

Increasing N_a leads to a further decrease of the electron density, until at $N_d = N_a$ the semiconductor becomes fully compensated, and the Fermi level returns to the middle of the energy gap. If N_a is larger than N_d, we have to consider the hole instead of the electron density

$$p = N_a - N_d. \tag{5.44}$$

In this case, the chemical potential is given by

$$\mu = E_V - k_B T \ln\left(\frac{N_a - N_d}{N_V}\right). \tag{5.45}$$

A p-type semiconductor of this kind is also partially compensated. The hole density is reduced due to the simultaneous presence of electrons from the donors.

5.4. Low-temperature regime for n-doped semiconductors

At low temperatures, the assumption of non-degenerate electrons cannot be sustained anymore. We will consider the case of a purely n-type semiconductor, i.e., $N_a = 0$. In this limit, the donors are only partially ionized. Therefore, the electron density is given by

$$n = N_d^+ + p, \tag{5.46}$$

where N_d^+ denotes the density of ionized donors, i.e.,

$$N_d^+ = N_d - N_d^0. \tag{5.47}$$

The density of neutral donors is given by

$$N_d^0 = \frac{N_d}{1 + \frac{1}{2}\exp[(E_d - \mu)/k_B T]}. \tag{5.48}$$

This occupation function can be derived when considering the following properties of a localized state such as a donor or acceptor level. Since there are two spin degenerate states, which will not be occupied at the same time due to the electron repulsion, a factor of 1/2 has to be introduced in front of the exponential function. Therefore, the difference between electron and hole density is given by

$$n - p = \frac{N_d}{2\exp[(\mu - E_d)/k_B T] + 1} . \tag{5.49}$$

The difference between the total donor density and the carrier density of electrons and holes can be written as

$$N_d - n + p = N_d \frac{2\exp[(\mu - E_d)/k_B T]}{2\exp[(\mu - E_d)/k_B T] + 1} . \tag{5.50}$$

Combining these two equations leads to the following expression

$$n - p = \frac{N_d - n + p}{2} \exp[(E_d - \mu)/k_B T] . \tag{5.51}$$

Substituting the chemical potential from Eq. 5.28 results in

$$(n - p)\, n = \frac{N_C}{2} (N_d - n + p) \exp(-E_b/k_B T) , \tag{5.52}$$

where $E_b = E_C - E_d$ denotes the donor binding energy. This is the central equation for determining the temperature dependence of the carrier density for a doped semiconductor. Using $p = n_i^2/n$, a cubic equation in n has to be solved. In the case that the intrinsic contribution is negligible, only a quadratic equation remains to be solved

$$n^2 + \frac{N_C'}{2} n - \frac{N_C' N_d}{2} = 0 , \tag{5.53}$$

where $N_C' = N_C \exp[-E_d/(k_B T)]$. The physical solution of this equation is

$$n = \frac{N_C'}{4} \left(-1 + \sqrt{1 + \frac{8N_d}{N_C'}} \right) . \tag{5.54}$$

At low temperatures, the intrinsic carrier density can be neglected and $N_d \gg N_C'$. Therefore, the carrier density at low temperatures follows from

$$n = \sqrt{\frac{N_C\, N_d}{2}} \exp\left(\frac{-E_b}{2\, k_B T}\right) . \tag{5.55}$$

The electron density increases exponentially with an activation energy e-qual to half of the donor binding energy. The chemical potential is obtained by replacing n in Eq. 5.55 with the expression from Eq. 5.28

$$\mu = E_G - \frac{E_b}{2} + \frac{k_B T}{2} \ln\left(\frac{N_d}{2\, N_C}\right) . \tag{5.56}$$

The chemical potential is located between the donor level and the conduction band edge. It decreases to lower energies with increasing temperature. At higher temperatures, when the exponential factor in N_C' approaches one, $N_d \ll N_C'$. However, we can still neglect the intrinsic carrier density. In this case, the square root in Eq. 5.54 can be expanded, and we recover the previous result $n = N_d$ for the saturation regime. At very high temperatures, the intrinsic regime is reproduced. In Tab. 5.6 the different regimes

Table 5.6. Temperature dependence of the electron concentration in an n-type semiconductor for the various temperature regimes.

Temperature	Low	Intermediate	High
Regime	Extrinsic (ionization)	Extrinsic (saturation)	Intrinsic
Prefactor	$\sqrt{N_C(T)\, N_d/2}$	N_d	$\sqrt{N_C(T)\, N_V(T)}$
Dominating factor	$\exp(\frac{-E_b}{2\, k_B T})$	1	$\exp(-\frac{E_G}{2\, k_B T})$
Chemical potential (Fermi level)	$\mu = E_d + \frac{E_b}{2} - \frac{k_B T}{2} \ln(\frac{2N_C(T)}{N_d})$	$\mu = E_C - k_B T \ln(\frac{N_C(T)}{N_d})$	$\mu = \frac{E_G}{2} - \frac{3k_B T}{4} \ln(\frac{m_e^*}{m_h^*})$

are listed including their respective temperature dependencies. The complete temperature dependence of n can be obtained by solving the cubic equation

$$n^3 + \frac{N_C'}{2} n^2 - \left(n_i^2 + \frac{N_C' N_d}{2} \right) n - \frac{N_C' n_i^2}{2} = 0 . \tag{5.57}$$

The temperature dependence of the chemical potential can be calculated by replacing n in Eq. 5.28 with the solution of the above cubic equation

$$\mu = E_C + k_B T \ln \left(\frac{n(T)}{N_C(T)} \right) . \tag{5.58}$$

In Fig. 5.5, the chemical potential and the logarithm of the electron concentration for a doped semiconductor are plotted as a function of temperature. The following values were used: $N_d = 1 \times 10^{17}$ cm^{-3}, $N_C(300\ K) = N_V(300\ K) = 5.2 \times 10^{18}$ cm^{-3}, $E_G = 0.66$ eV, and $E_d = 10$ meV. At low

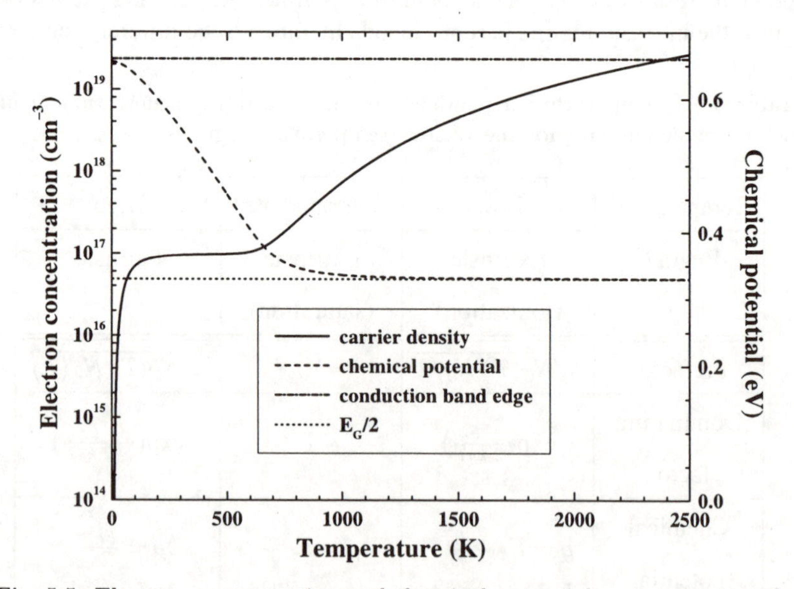

Fig. 5.5. Electron concentration and chemical potential vs temperature for an n-type semiconductor. The parameters are given in the text.

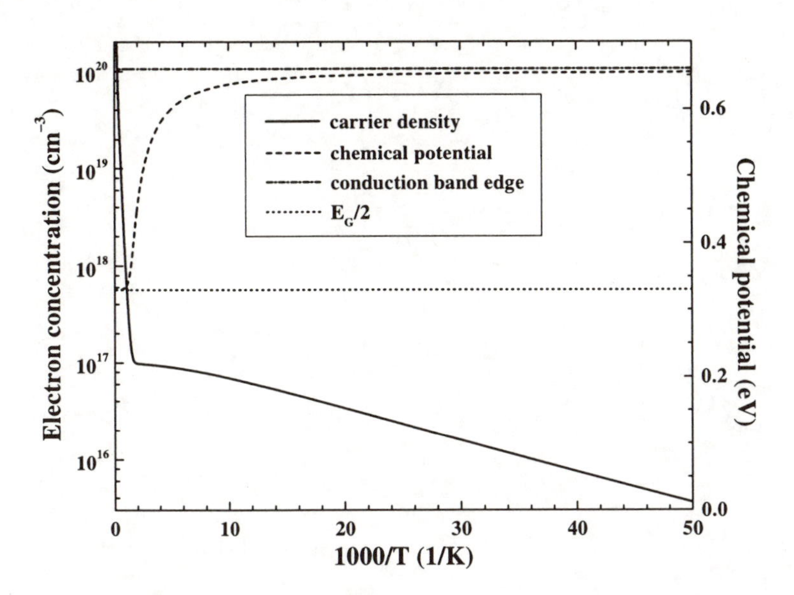

Fig. 5.6. Electron concentration and chemical potential vs inverse temperature for an n-type semiconductor. The parameters are given in the text.

temperatures, the carrier density is strongly reduced (carrier freeze out). The saturation regime with a carrier density equal to the doping density is clearly visible between 150 and 600 K. For higher temperatures, the intrinsic carrier density dominates. The chemical potential is at $E_G - E_b/2$ for very low temperatures. In the saturation regime, it moves from this value down to the middle of the energy gap, where it remains for high temperatures. In Fig. 5.6 the chemical potential and the logarithm of the electron concentration are plotted versus inverse temperature. The slope of the solid line approaches at high temperatures $E_G/2$, while at low temperatures it corresponds to $E_d/2$. It is obvious from Fig. 5.6 that the chemical potential remains near the conduction band edge ($E_F = E_G - E_b/2$ at 0 K) at low temperatures.

CHAPTER 6
CARRIER TRANSPORT

The basic equations for transport were already given in Chapter 4. If a constant electric field is applied to a semiconductor with an isotropic effective mass, we arrive using Eqs. 4.9 and 4.13 at the following equation for the time evolution of the crystal momentum $\underline{p} = \hbar \underline{k}$

$$\frac{d\underline{k}}{dt} = -\frac{e}{\hbar} \underline{F_E} , \tag{6.1}$$

where $\underline{F_E}$ denotes the strength of the applied electric field. Integrating this equation with the initial condition $\underline{k}(t = 0) = 0$ leads to

$$\underline{k}(t) = -\frac{e}{\hbar} \underline{F_E}\, t . \tag{6.2}$$

If there were no scattering processes, the drift velocity $\langle \underline{v} \rangle = \hbar \underline{k}/m^*$ would increase continuously with time. However, elastic and inelastic scattering processes result in a change of the direction and value of $\underline{k}$, respectively, so that the time t is replaced by an average scattering time $\langle \tau \rangle$. The resulting average drift velocity $\langle \underline{v}_e \rangle$ has the following form

$$\langle \underline{v}_e \rangle = -\frac{e \langle \tau \rangle}{m_e^*} \underline{F_E} . \tag{6.3}$$

The proportionality factor between the drift velocity and the electric field strength is called the electron mobility μ_e

$$\mu_e = \frac{e \langle \tau \rangle}{m_e^*} . \tag{6.4}$$

The average scattering time $\langle \tau \rangle$ can be used to calculate the mean free path $\langle l \rangle = \langle v \rangle \langle \tau \rangle$, which corresponds to the average distance between two scattering events.

The current density for electrons is related to the carrier density and drift velocity by $\underline{j}_e = -en\langle \underline{v}_e \rangle$. Therefore,

$$\underline{j}_e = \frac{e^2\, n\, \langle \tau \rangle}{m_e^*}\, \underline{F}_E = e\, n\, \mu_e\, \underline{F}_E \,. \tag{6.5}$$

The proportionality factor between the current density and the electric field strength is called the electron conductivity σ_e, which is the inverse of the resistivity ρ_e,

$$\sigma_e = \frac{1}{\rho_e} = e\, n\, \mu_e = \frac{e^2\, n\, \langle \tau \rangle}{m_e^*} \,. \tag{6.6}$$

For an anisotropic effective mass, the conductivity becomes a tensor. The drift velocity for holes in the valence band is given by

$$\langle \underline{v}_h \rangle = \frac{e\, \langle \tau \rangle}{m_h^*}\, \underline{F}_E \,. \tag{6.7}$$

Therefore, the drift velocity of electrons and holes have opposite signs. The current density for holes is given by $\underline{j}_h = ep\langle \underline{v}_h \rangle$. Since electrons and holes have opposite charge and opposite drift velocities, the total current density is therefore obtained by adding the electron and hole current density $\underline{j}_e + \underline{j}_h$. The total current density is proportional to the applied electric field. The total conductivity σ_t is therefore given by

$$\sigma_t = \sigma_e + \sigma_h = e\, (n\, \mu_e + p\, \mu_h) \,. \tag{6.8}$$

1. The Drude Model

Long before semiconductors became important, Drude formulated a simple theory of the electron transport in conductors. The main assumption of the Drude model is that carriers are scattered, i.e., their momentum is changed, on an average time scale τ. The resulting transport equation has the simple form

$$\frac{d\underline{p}}{dt} = -e\, \underline{F}_E - \frac{\underline{p}}{\tau} \,. \tag{6.9}$$

The steady state solution to this equation is

$$\underline{p} \;=\; -e\,\tau\,\underline{F}_E\,. \tag{6.10}$$

The average momentum or velocity is then given by

$$\langle \underline{v} \rangle \;=\; -\frac{e\,\langle\tau\rangle}{m^*}\,\underline{F}_E\,. \tag{6.11}$$

Note that we have derived this equation without any assumptions on the quantum mechanical level. The only input we have used is the effective mass. This result is identical to Eq. 6.3. The conductivity is therefore the same as in Eq. 6.6. The conductivity of a semiconductor can therefore be determined only from the carrier density, which was already discussed in the previous chapter, the effective mass, and the scattering time. However, as we will show in the next section, the determination of the scattering time is not an easy task. In Chapter 8, we will discuss different scattering mechanisms and summarize their temperature dependence.

2. The Boltzmann Equation

In the previous chapter, we only dealt with carriers at thermal equilibrium. However, when an electric field is applied, the situation is changed, since the energy distribution of the carriers is also affected. The scattering time itself may depend on energy. In order to determine, how the distribution function is influenced by an electric field, the Boltzmann transport equation has to be considered. Let us assume an electron at the position $\underline{r}$ with momentum $\underline{k}$ at time t. The Boltzmann equation for the distribution function $f(\underline{k},\underline{r},t)$ (cf. Eq. 5.24) can be derived by stating that the complete change in the distribution function df/dt is only caused by collisions, i.e.,

$$\frac{df}{dt} = \frac{\partial f}{\partial \underline{r}}\cdot\frac{d\underline{r}}{dt} + \frac{\partial f}{\partial \underline{k}}\cdot\frac{d\underline{k}}{dt} + \frac{\partial f}{\partial t} = \left(\frac{\partial f}{\partial t}\right)_{coll}\,. \tag{6.12}$$

The left-hand-side of Eq. 6.12 is the complete differential of the distribution function, while the right-hand-side denotes the variation of the particle number by collisions. Using the equation of motion in reciprocal space (Eq. 6.1), we arrive at the Boltzmann equation for electron transport in a semiconductor

$$\underline{\mathrm{v}} \cdot \underline{\nabla}_{r} f - \frac{e}{\hbar} \underline{F}_E \cdot \underline{\nabla}_{k} f + \frac{\partial f}{\partial t} = \left(\frac{\partial f}{\partial t}\right)_{coll} . \tag{6.13}$$

The first term describes diffusion processes, since it requires a change of f in real space. The second term shows the variation of f under the application of an external electric field. The third term gives the variation due to the explicit time dependence. This term will not be considered in the following. The most difficult term is the collision term. One often uses the so-called relaxation-time approximation for the collision term, which can be expressed as

$$\left(\frac{\partial f}{\partial t}\right)_{coll} = -\frac{f - f_0}{\tau(\underline{k})} . \tag{6.14}$$

Here, f_0 denotes the equilibrium distribution function. Assuming that the distribution function has no explicit time dependence, Eq. 6.13 reduces to

$$\underline{\mathrm{v}} \cdot \underline{\nabla}_{r} f - \frac{e}{\hbar} \underline{F}_E \cdot \underline{\nabla}_{k} f = -\frac{f - f_0}{\tau(\underline{k})} . \tag{6.15}$$

If f does not depend on the spatial coordinate $\underline{r}$, Eq. 6.15 can be further simplified. The deviation from the equilibrium distribution function is then given by

$$f - f_0 = \frac{e}{\hbar} \tau(\underline{k}) \, \underline{F}_E \cdot \underline{\nabla}_{k} f(\underline{k}) . \tag{6.16}$$

Assuming that the deviation is small, we can replace on the right-hand-side f by f_0, which depends only through $E(\underline{k})$ on the wavevector

$$f = f_0 + \frac{e}{\hbar} \tau(\underline{k}) \frac{\partial f_0}{\partial E} \underline{F}_E \cdot \underline{\nabla}_{k} E(\underline{k}) . \tag{6.17}$$

The last term in Eq. 6.17 is the group velocity of the electron. Taking the electric field parallel to the z-direction, we arrive at

$$f = f_0 + e \, \tau(\underline{k}) \, F_E \, \mathrm{v}_z \frac{\partial f_0}{\partial E} . \tag{6.18}$$

The average drift velocity is calculated using the distribution function f by evaluating the following integrals

$$\langle \mathrm{v}_z \rangle \quad = \quad \frac{\int d^3k \, \mathrm{v}_z(\underline{k}) \, f(\underline{k})}{\int d^3k \, f(\underline{k})} \, . \tag{6.19}$$

For crystals with isotropic bands, the integral over f_0 in the numerator of Eq. 6.19 vanishes, while in the denominator the integral over the correction term vanishes. We therefore arrive at the following expression for the current density of electrons in a semiconductor

$$j_z(T, F_E) \quad = \quad -e^2 \, n \, F_E \, \frac{\int d^3k \, \tau(\underline{k}) \, \mathrm{v}_z^2(\underline{k}) \, \frac{\partial f_0}{\partial E}(\underline{k})}{\int d^3k \, f_0} \, . \tag{6.20}$$

Assuming that the electrons obey Maxwell-Boltzmann statistics, we can write $f_0 = const. \exp(-E/k_B T)$. In this case, the derivative with respect to the energy becomes

$$\frac{\partial f_0}{\partial E} \quad = \quad -\frac{1}{k_B T} \, f_0 \, . \tag{6.21}$$

Using a parabolic dispersion, i.e., the effective mass picture, and the density of states, we finally arrive at the following expression for the mobility

$$\mu_e(T) \quad = \quad \frac{2 \, m_e^*}{3} \, \frac{e}{k_B T} \, \frac{\int_0^\infty dE \, E^{3/2} \, \tau(E) \, f_0(E)}{\int_0^\infty dE \, E^{1/2} \, f_0(E)} \, . \tag{6.22}$$

Noting that

$$\int_0^\infty dE \, E^{3/2} \exp(-E/k_B T) = \frac{3 \, k_B T}{2} \int_0^\infty dE \, E^{1/2} \exp(-E/k_B T) \, , \tag{6.23}$$

the average scattering time is determined by

$$\langle \tau(T) \rangle \quad = \quad \frac{\int_0^\infty dE \, E^{3/2} \, \tau(E) \, f_0(E)}{\int_0^\infty dE \, E^{3/2} \, f_0(E)} \, . \tag{6.24}$$

The average scattering time is therefore obtained by using $E^{3/2} \, f_0$ as the weighting function. One can extend the treatment to degenerate electrons, which obey Fermi-Dirac statistics. In order to determine the mobility, one needs to know the energy dependence of the scattering time. Using Eq. 6.24, one can then calculate the average scattering time, which is usually not an easy task, and derive an expression for the temperature dependence of the mobility. We will return to this subject in Chapter 8 after we have discussed the phonon modes of the crystal, which are important for some scattering mechanisms.

3. The Hall Effect

In the previous sections, we considered the effect of an electric field on the carrier transport. A more general approach includes also the effect of a magnetic field. Transport under the simultaneous presence of an electric and a magnetic field perpendicular to it is usually referred to as the Hall effect. This effect is very important for semiconductors, since it allows the determination of the carrier density as well as of the type of carrier, i.e., electron or hole. Furthermore, information on the scattering time can also be obtained. Assuming that the electric field component F_E is parallel to the x-axis, while the magnetic field B is parallel to the z-axis, the electrons will drift parallel to the x-direction and simultaneously be deflected in the y-direction by the magnetic field. Since there is no current flow in the y-direction, an electric field will build up in this direction, which will cancel the Lorentz force. This field is called the Hall field. The configuration is shown schematically in Fig. 6.1. Following Eq. 6.9, the equation of motion in an electric and a magnetic field can be expressed as

$$\frac{d\underline{p}}{dt} = -e\left(\underline{F}_E + \frac{1}{m^*}\,\underline{p}\times\underline{B}\right) - \frac{\underline{p}}{\tau}\,. \tag{6.25}$$

In steady state, the time derivative is zero. Since $\underline{B} = B\,\hat{z}$, the following relations can be deduced

$$p_x = -e\,\tau\,F_x - \frac{e\,\tau}{m^*}\,p_y\,B\,, \tag{6.26}$$

$$p_y = -e\,\tau\,F_y + \frac{e\,\tau}{m^*}\,p_x\,B\,, \tag{6.27}$$

$$p_z = 0\,. \tag{6.28}$$

Since the current density is related to the drift velocity and, therefore, to the momentum by $\underline{j} = (en/m^*)\underline{p}$, we can rewrite the first two equations

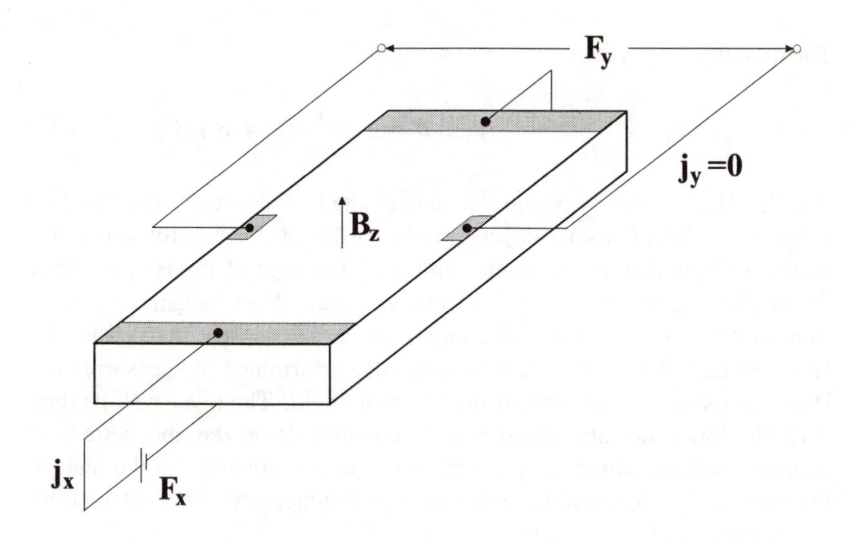

Fig. 6.1. Schematic diagram of the Hall effect.

describing the motion in the x-y plane in matrix form

$$\begin{pmatrix} 1 & \mu B \\ -\mu B & 1 \end{pmatrix} \begin{pmatrix} j_x \\ j_y \end{pmatrix} = \sigma_0 \begin{pmatrix} F_x \\ F_y \end{pmatrix}, \tag{6.29}$$

where $\sigma_0 = e\, n\, \mu$, or by using the definition $\rho_0 = 1/\sigma_0$

$$\rho_0 \begin{pmatrix} 1 & \mu B \\ -\mu B & 1 \end{pmatrix} \begin{pmatrix} j_x \\ j_y \end{pmatrix} = \begin{pmatrix} F_x \\ F_y \end{pmatrix}. \tag{6.30}$$

ρ_0 denotes the resistivity for vanishing magnetic field. It follows that in a magnetic field the conductivity (the resistivity) is not scalar quantity, but a two-dimensional tensor. The conductivity is obtained by inverting Eq. 6.30

$$\begin{pmatrix} j_x \\ j_y \end{pmatrix} = \frac{\sigma_0}{1 + \mu^2 B^2} \begin{pmatrix} 1 & -\mu B \\ \mu B & 1 \end{pmatrix} \begin{pmatrix} F_x \\ F_y \end{pmatrix}. \tag{6.31}$$

where $\sigma_0 = 1/\rho_0$. From Eq. 6.30, the Hall field can be determined using the condition $j_y = 0$

$$F_y = -\rho_0\, \mu\, B\, j_x\,. \tag{6.32}$$

The quantity

$$\rho_H = \frac{F_y}{j_x} = -\rho_0 \mu B = -\frac{1}{e\,n} B = R_H B \tag{6.33}$$

is called Hall resistivity, while R_H denotes the Hall constant. The absolute value of the Hall constant depends only on the carrier density, it is completely independent of the scattering time. The sign of the Hall constant determines the carrier type, i.e., electron or hole. R_H is negative for electrons and positive for holes. The larger the carrier density, the smaller the Hall constant. The carrier density is usually determined by measuring the Hall resistivity as a function of the magnetic field. The inverse slope then gives the carrier density. Measuring the conductivity at zero magnetic field leads to a determination of the mobility μ and, consequently, of the scattering time τ. The longitudinal resistance or conductance is not affected by the magnetic field.

If the current in the semiconductor is determined by both electron and hole transport, the calculation becomes somewhat more involved. The resulting Hall constant is of the form

$$R_H = \frac{p\,\mu_h^2 - n\,\mu_e^2}{e\,(p\,\mu_h + n\,\mu_e)^2} . \tag{6.34}$$

While for a single carrier type the longitudinal resistance is not affected by the magnetic field, the situation changes, when two different types of carriers are present and a longitudinal magnetoresistance appears. This magnetoresistance depends quadratically on the magnetic field. By measuring the resistance at $B = 0$, the Hall constant R_H, and the magnetoresistance, the four unknown quantities in Eq. 6.34, i.e., p, n, μ_n, and μ_p, cannot be determined. If the electron and hole densities are equal such as for an intrinsic semiconductor or a semimetal, Eq. 6.34 reduces to

$$R_H = \frac{\mu_h^2 - \mu_e^2}{(\mu_h + \mu_e)^2} \frac{1}{e\,n} . \tag{6.35}$$

In this case there are only three unknown quantities. For $\mu_h = \mu_e$, the Hall constant becomes zero, but not the conductance and the magnetoresistance.

4. Mobilities of Selected Semiconductors

In this chapter, it has become clear that the mobility is one of the key quantities in semiconductor physics. It is proportional to the average scattering time and inversely proportional to the effective mass. In Tab. 6.1 the room temperature electron and hole mobilities for a number of semiconductors are listed. Figure 6.2 displays the electron mobility as a function of the inverse effective mass. Recent advances in fabrication techniques have resulted in an improved mobility for some materials so that some values may have changed. In some cases, the impurity content has also been taken into account. Nevertheless, the mobility of the group IV semiconductors remains almost constant, while there appears to be an overall increase of the mobility with decreasing effective mass. This indicates that the scattering process, which determines the average scattering time at room temperature, is very similar for most semiconductors. Typical scattering times are in the femtosecond time range so that these scattering processes are extremely fast at room temperature. The typical mean free path is in the

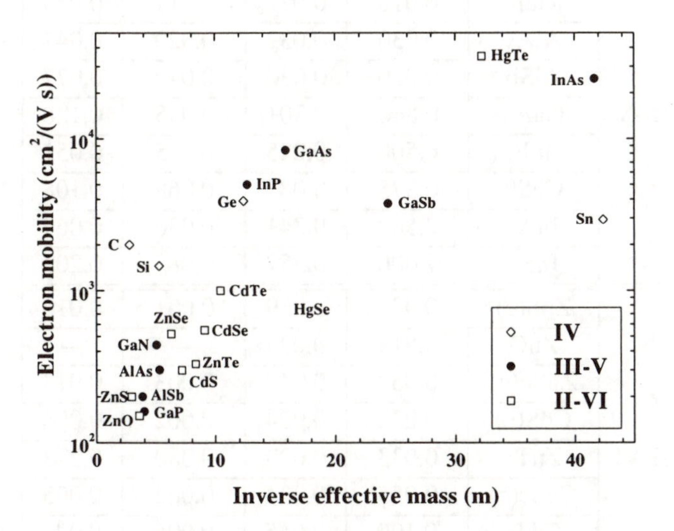

Fig. 6.2. Electron mobility vs inverse effective mass for a number of group IV, III-V, and II-VI semiconductors at room temperature.

range of several 10 nm assuming a velocity of the order of 10^5 ms^{-1}. For two-dimensional electron gases in GaAs at low temperatures, mobilities exceeding 1000 m^2(Vs)$^{-1}$ have been achieved corresponding to a scattering time beyond 360 ps. In this case, the mean free path can reach values of several tens of μm.

Table 6.1. Electron and hole mobilities at room temperature of selected group IV, III-V, and II-VI semiconductors. The mobility is usually given in cm^2(Vs)$^{-1}$, which corresponds to 10^{-4} T^{-1} [1 T$^{-1}\hat{=}$ 1 m^2(Vs)$^{-1}$].

Type	Material	μ_e (T^{-1})	τ_e (ps)	μ_h (T^{-1})	τ_h (ps)
IV	Diamond	0.20	0.41	0.21	1.29
	Si	0.145	0.16	0.037	0.113
	Ge	0.390	0.18	0.180	0.291
	Sn	0.294	0.039	0.299	0.332
III-V	GaN	0.044	0.050	—	—
	GaP	0.016	0.027	0.014	0.053
	AlAs	0.030	0.032	0.020	0.047
	AlSb	0.020	0.030	0.040	0.077
	GaAs	0.840	0.301	0.045	0.128
	InP	0.500	0.225	0.015	0.051
	GaSb	0.375	0.087	0.068	0.108
	InAs	2.500	0.341	0.030	0.069
	InSb	7.000	0.557	0.085	0.203
II-VI	ZnS(c)	0.020	0.039	0.001	0.010
	ZnO	0.015	0.024	—	—
	ZnSe(c)	0.052	0.047	0.003	0.013
	CdS(c)	0.030	0.024	0.002	0.006
	ZnTe	0.033	0.023	0.080	0.273
	CdSe(c)	0.055	0.034	0.002	0.005
	CdTe	0.100	0.055	0.006	0.022
	HgSe	1.500	—	—	—
	HgTe	3.500	0.617	—	—

CHAPTER 7

PHONONS AND PHONON STATISTICS

So far we have completely neglected that the core of the atoms forming the lattice are not fixed in space, but actually oscillate around their equilibrium positions. This effect becomes increasingly important as the temperature rises from very low temperatures to room temperature. These lattice vibrations are also important in connection with transport processes, since they can scatter electrons. The quantized version of these lattice vibrations are called phonons, which are described by a vector field. In contrast to electrons, which are Fermions, phonons are spin-1 particles so that they obey Bose-Einstein statistics. After discussing the classical harmonic spectrum of lattice vibrations, we will look at the density of states and the statistics of phonons. This allows us to determine the number of thermally excited phonons at any temperature.

1. Acoustic Phonons

The simplest solid is a monatomic linear chain in one dimension. For this system, there is only one atom per unit cell. We will assume a restoring force proportional to the displacement and consider only the interaction between nearest neighbors. The equation of motion for the displacement $u(na,t)$, where a denotes the lattice constant, n the position of the nth atom, and M the atomic mass, is then given by

$$\begin{aligned} M\,\frac{d^2u(na,t)}{dt^2} &= -\kappa\,[u(na,t) - u[(n+1)a,t]] \\ &\quad -\kappa\,[u(na,t) - u[(n-1)a,t]]\ , \end{aligned} \qquad (7.1)$$

where κ denotes the spring constant of the restoring force. This equation can be rewritten by rearranging its right-hand side to yield

$$M \frac{d^2u(na,t)}{dt^2} = \\ -\kappa \, [2 \, u(na,t) \; - \; u[(n+1)a,t] \; - \; u[(n-1)a,t]] \; . \tag{7.2}$$

In analogy to the solution of the Schrödinger equation using Bloch functions, the solution to this equation is most easily obtained using the function

$$u(na,t) \;=\; A \, e^{ikna} \, e^{i\omega t} \; , \tag{7.3}$$

where A is determined by the initial condition. Applying periodic boundary conditions, i.e.,

$$u[(N+1)a,t] \;=\; u(a,t) \, , \tag{7.4}$$

the following values for k are obtained

$$k \;=\; \frac{2\,\pi}{a} \frac{n}{N} \, . \tag{7.5}$$

By substituting this information in Eq. 7.1, a simple equation for the possible frequencies is derived

$$-\omega^2 \, M \;=\; -2 \, \kappa [1 \; - \; \cos(ka)] \; . \tag{7.6}$$

The solution of this equation, which is shown in Fig. 7.1, is given by

$$\omega(k) \;=\; 2 \sqrt{\frac{\kappa}{M}} \left| \sin\left(\frac{k\,a}{2} \right) \right| \, . \tag{7.7}$$

In the low-frequency limit, i.e., $k \ll \pi/a$, the sine function on the right-hand side can be expanded as

$$\omega(k) \;=\; \sqrt{\frac{\kappa}{M}} \, a \, |k| \; . \tag{7.8}$$

This dispersion relation constitutes the linear relationship between frequency and wavevector for sound waves. In fact, in the continuum limit, Eq. 7.1 transforms into

$$\rho \frac{\partial^2 u(x,t)}{\partial t^2} \;=\; C \frac{\partial^2 u(x,t)}{\partial x^2} \, , \tag{7.9}$$

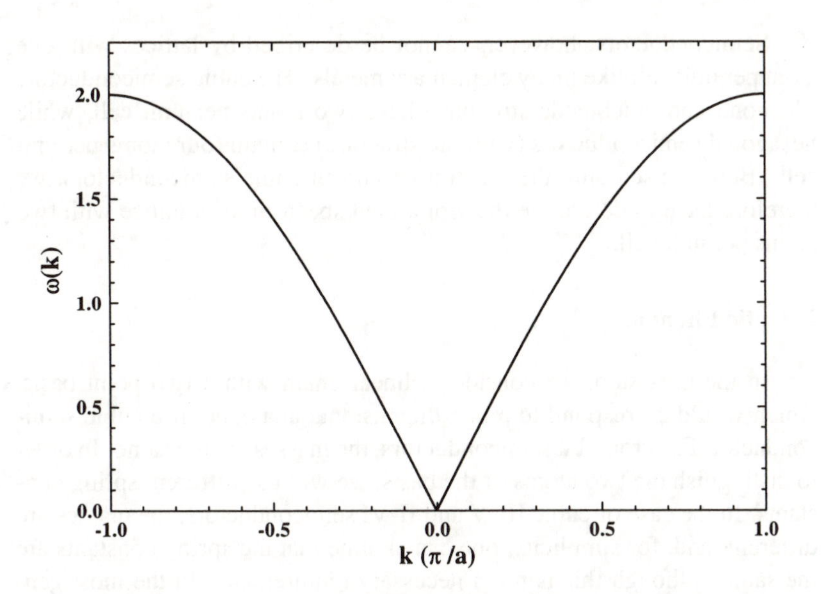

Fig. 7.1. Frequency spectrum of lattice vibrations in the first Brillouin zone for a monoatomic lattice. The frequency scale is given in units of $\sqrt{\kappa/M}$.

where ρ denotes the density and C the elastic constant. In this limit the dispersion relation is given by

$$\omega(k) \;=\; v_s\,|k| \tag{7.10}$$

with $v_s = \sqrt{C/\rho}$ denoting the sound velocity. A comparison of Eq. 7.10 with Eq. 7.8 leads to a relation between the microscopic quantity κ and the macroscopic quantity C using $\rho = M/a^3$

$$\kappa \;=\; C\,a\,. \tag{7.11}$$

In three dimensions, a total of three acoustic branches is obtained, one longitudinal (LA) branch and two transverse (TA) branches. Since the speed of sound for longitudinal waves is usually higher than that for transverse waves, the longitudinal branch has the highest frequency at the zone boundary with a value of 6.6 THz (corresponding to 220 cm^{-1} or 27.3 meV) in GaAs.

Semiconductors, however, cannot be described by lattices with one atom per unit cell like many elementary metals. The cubic semiconductors (diamond and zincblende structure) have two atoms per unit cell, while hexagonal semiconductors (wurtzite structure) contain four atoms per unit cell. Before discussing the phonon spectra of some semiconductors, we therefore have to determine the vibrational spectrum of a lattice with two atoms per unit cell.

2. Optic Phonons

In the next step, we consider a linear chain with a two point basis, which would correspond to a one-dimensional analogue of a cubic semiconductor. For group IV semiconductors, the masses are the same. In order to distinguish the two atoms of the basis, we will use different spring constants. In the case of cubic III-V and II-VI semiconductors, the masses are different, and, for simplicity, one can assume that the spring constants are the same, although this is not a necessary requirement. In the most general case, we assume different masses and spring constants. In the following, the displacement of the atom with mass M_1 (M_2) is denoted $u_1(na,t)$ ($u_2(na,t)$). For nearest neighbor interaction, the equations for the oscillation of the atoms around their equilibrium position are then of the form

$$M_1 \frac{d^2u_1(na,t)}{dt^2} = -\kappa_1 \left[u_1(na,t) - u_2(na,t)\right] -\kappa_2 \left[u_1(na,t) - u_2[(n-1)a,t]\right] , \quad (7.12)$$

$$M_2 \frac{d^2u_2(na,t)}{dt^2} = -\kappa_1 \left[u_2(na,t) - u_1(na,t)\right] -\kappa_2 \left[u_2(na,t) - u_1((n+1)a,t)\right] . \quad (7.13)$$

We again use a similar function as in the case of acoustic phonons, i.e.,

$$u_1(na,t) = \varepsilon_1 \, e^{ikna} \, e^{i\omega t} ,$$
$$u_2(na,t) = \varepsilon_2 \, e^{ikna} \, e^{i\omega t} , \quad (7.14)$$

where ε_1 and ε_2 are again determined by the initial conditions. We will also assume periodic boundary conditions, i.e.,

$$u_1[(N+1)a,t] = u_1(a,t) ,$$
$$u_2[(N+1)a,t] = u_2(a,t) . \quad (7.15)$$

Using the function of Eq. 7.14, we obtain the following equation in matrix form, which determines the frequencies,

$$\begin{pmatrix} \kappa_1+\kappa_2-\omega^2 M_1 & -\kappa_1-\kappa_2\, e^{-ika} \\ -\kappa_1-\kappa_2\, e^{ika} & \kappa_1+\kappa_2-\omega^2 M_2 \end{pmatrix} \begin{pmatrix} \varepsilon_1 \\ \varepsilon_2 \end{pmatrix} = 0\,. \tag{7.16}$$

This equation has non-trivial solutions only, if the determinant of the coefficient matrix vanishes. The eigenvalues are therefore obtained from

$$\omega^4 - \omega^2\,(\kappa_1+\kappa_2)\,\frac{M_1+M_2}{M_1\,M_2} + 2\,\frac{\kappa_1\,\kappa_2}{M_1\,M_2}\,[1 - \cos(k\,a)] = 0\,. \tag{7.17}$$

The solutions to this equation are given by

$$\omega_{1,2}^2(k) = G_1\left[1 \pm \sqrt{1 - G_2\,\sin^2(\frac{k\,a}{2})}\right] \tag{7.18}$$

with

$$G_1(\kappa_1,\kappa_2,M_1,M_2) = \frac{\kappa_1+\kappa_2}{2}\,\frac{M_1+M_2}{M_1\,M_2} \tag{7.19}$$

and

$$G_2(\kappa_1,\kappa_2,M_1,M_2) = \frac{16\,\kappa_1\,\kappa_2}{(\kappa_1+\kappa_2)^2}\,\frac{M_1\,M_2}{(M_1+M_2)^2}\,. \tag{7.20}$$

In Fig. 7.2, both solutions $\omega_1(k)$ and $\omega_2(k)$ are shown for $\kappa_1 = \kappa_2$ and $M_1 < M_2$. The lower branch looks very similar to the acoustic branch of Fig. 7.1. If the expression for the low-frequency branch (minus sign) is expanded for small values of k, one obtains

$$\omega_1^2(k) = \frac{\kappa_1\,\kappa_2}{\kappa_1+\kappa_2}\,\frac{a^2}{M_1+M_2}\,k^2\,, \tag{7.21}$$

which again is the dispersion relation for sound waves. However, the second branch has a very different behavior for small values of k

$$\omega_2^2(k) = (\kappa_1+\kappa_2)\,\frac{M_1+M_2}{M_1\,M_2} - \frac{\kappa_1\,\kappa_2}{\kappa_1+\kappa_2}\,\frac{a^2}{M_1+M_2}\,k^2\,. \tag{7.22}$$

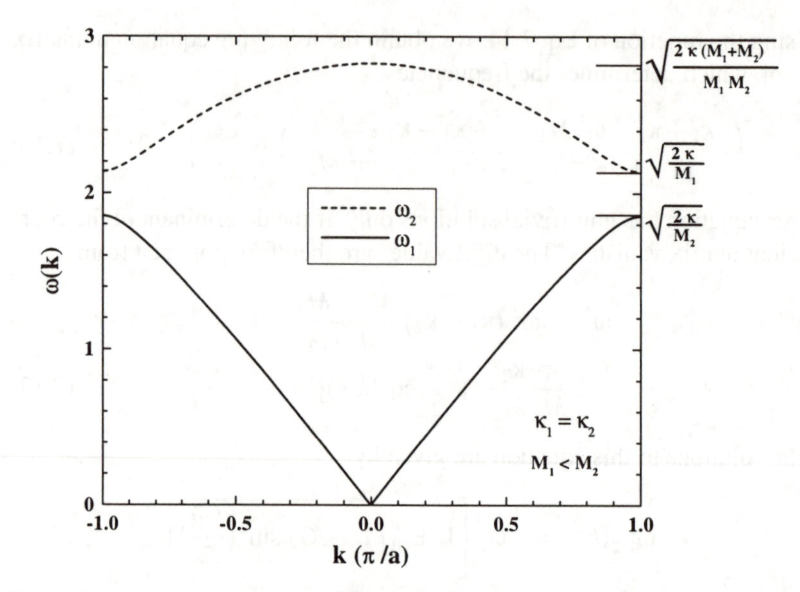

Fig. 7.2. Frequency spectrum of lattice vibrations in the first Brillouin zone for a lattice with a two point basis. The frequency scale is given in units of $\sqrt{G_1/2}$.

In contrast to the acoustic branch, the frequency of this branch does not vanish for $k = 0$, but achieves a finite value of

$$\omega_2(k=0) \quad = \quad \sqrt{(\kappa_1 + \kappa_2)\,\frac{M_1 + M_2}{M_1\,M_2}}\,. \tag{7.23}$$

This branch is called the optic branch. If we now solve Eq. 7.16 in the limit $k \to 0$, we find that for the acoustic branch ($\omega \to 0$) the atoms are oscillating in phase, i.e.,

$$\varepsilon_1 \quad = \quad \varepsilon_2\,, \tag{7.24}$$

while for the optic branch ($\omega \neq 0$) the atoms within one unit cell oscillate with opposite phases, i.e.,

$$\varepsilon_1 \quad = \quad -\frac{M_2}{M_1}\,\varepsilon_2\,. \tag{7.25}$$

Since in polar semiconductors two atoms with different masses oscillating 180° out of phase can produce a dipole moment, this mode can couple to electromagnetic radiation. Therefore, this branch is usually referred to as the optic branch.

For the elementary semiconductors such as Si and Ge, the masses are equal ($M_1 = M_2 = M$), and the following frequencies at the zone center and boundary are obtained by assuming $\kappa_2 > \kappa_1$

$$\begin{array}{lll} k=0: & \omega_1 = 0 & \omega_2 = \sqrt{\dfrac{2(\kappa_1+\kappa_2)}{M}} \\ k=\dfrac{\pi}{a}: & \omega_1 = \sqrt{2\,\dfrac{\kappa_1}{M}} & \omega_2 = \sqrt{2\,\dfrac{\kappa_2}{M}} \end{array} \quad . \tag{7.26}$$

From these equations, it follows that a material with heavier atoms should exhibit a smaller optic phonon frequency. Comparing Si and Ge, we note that the energy ($\hbar\,\omega$) of the zone center optic phonon for Si is 64.1 meV, while for Ge it is 37.7 meV, so that the ratio of the energies is 1.70. According to Eq. 7.26, the energies should scale with a factor of 1.61 assuming that the spring constants are the same. The trend that the optic phonon frequencies decrease with increasing mass is generally observed.

In the case of cubic III-V and II-VI semiconductors, the masses are different, and we assume that the spring constant are equal ($\kappa_1 = \kappa_2 = \kappa$). We therefore obtain assuming $M_1 < M_2$ at the zone center and boundary the following frequencies

$$\begin{array}{lll} k=0: & \omega_1 = 0 & \omega_2 = \sqrt{2\,\kappa\,\dfrac{M_1+M_2}{M_1\,M_2}} \\ k=\dfrac{\pi}{a}: & \omega_1 = \sqrt{2\,\dfrac{\kappa}{M_2}} & \omega_2 = \sqrt{2\,\dfrac{\kappa}{M_1}} \end{array} \quad . \tag{7.27}$$

In three dimensions, cubic semiconductors (diamond and zincblende structure) have three acoustic branches (1 LA, 2 TA) and three optic branches, one longitudinal (LO) and two transverse (TO) ones. Hexagonal semiconductors (wurtzite structure) contain four atoms per unit cell. These materials have therefore nine optic branches in addition to the three acoustic branches. Usually the LO branch at the zone center has the highest frequency. In Fig. 7.3, the phonon dispersion curves of Si (top) and GaAs

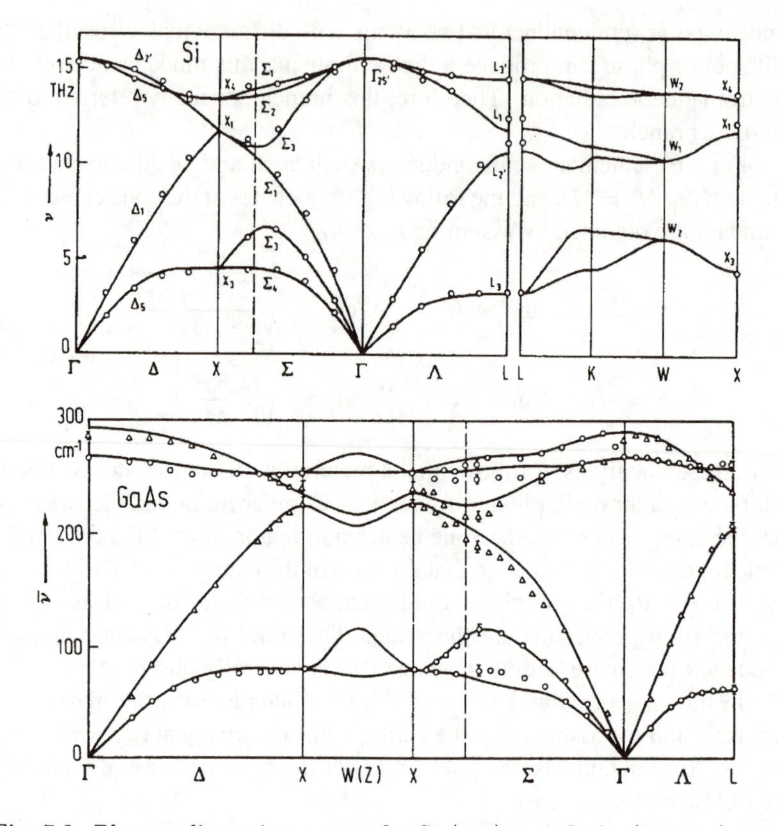

Fig. 7.3. Phonon dispersion curves for Si (top) and GaAs (bottom) taken from *Landolt-Börnstein, Series III, Vol. 22a.* The points refer to experimental data, the solid lines to the calculated spectra.

(bottom) are shown for several directions in reciprocal space. For some directions, the TA and TO modes are degenerate, for other directions all three modes of one branch have different frequencies. In Tab. 7.1, the largest optic phonon energy of many semiconductors is compiled together with their reduced atomic mass. In each group of semiconductors, the materials are listed in order of decreasing phonon frequency. The dependence on the reduced atomic mass can directly be seen for the group IV semiconductors. In Fig. 7.4, the optic phonon energies listed in Tab. 7.1 are plotted

Table 7.1. Largest optic phonon frequency at room temperature for selected group IV, III-V, and II-VI semiconductors (1 THz corresponds to 33.356 cm^{-1} or 4.136 meV). The reduced atomic mass is listed in the last column in units of u (1 u $\hat{=}$ 1.66054 $\times$ 10^{-27} kg).

Type	Material	$\hbar\omega_{LO}$ (THz)	$\hbar\omega_{LO}$ (cm^{-1})	$\hbar\omega_{LO}$ (meV)	$\frac{M_1 M_2}{M_1+M_2}$ (u)
IV	Diamond	39.9	1331	165	6.0
	Si	15.5	517	64.1	14.0
	Ge	9.0	300	37.2	36.3
	Sn	6.0	200	24.8	59.4
III-V	BN	39.1	1304	162	6.1
	AlN	26.8	894	111	9.2
	BP	24.9	830	103	8.0
	GaN	22.3	744	92.2	11.7
	AlP	15.0	500	62.0	14.4
	GaP	12.3	410	50.9	21.4
	AlAs	12.1	404	50.0	19.8
	InP	10.3	344	42.6	24.4
	AlSb	10.2	340	42.2	22.1
	GaAs	8.8	294	36.4	36.1
	InAs	7.2	240	29.8	45.3
	GaSb	6.35	212	26.3	44.3
	InSb	5.9	197	24.4	59.1
II-VI	ZnO	17.6	587	72.8	12.9
	ZnS(c)	10.5	350	43.4	21.5
	CdS(h)	9.1	302	37.6	25.0
	ZnSe(c)	7.59	253	31.4	35.8
	CdSe(h)	6.3	210	26.1	46.4
	ZnTe	6.2	207	25.6	43.2
	CdTe	5.08	169	21.0	59.8
	HgSe	4.9	163	20.3	56.7
	HgTe	4.4	147	18.2	78.0

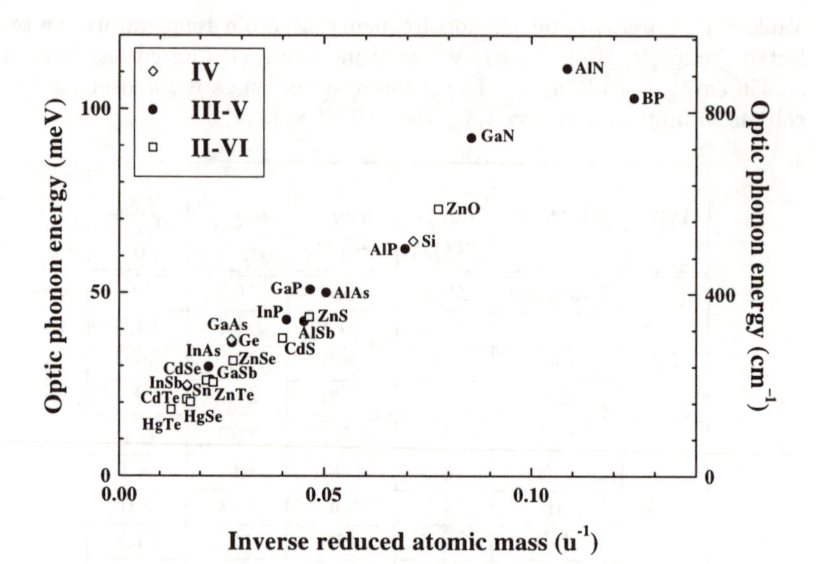

Fig. 7.4. Optic phonon energy vs inverse reduced atomic mass for group IV, III-V, and II-VI semiconductors. Diamond and BN also follow the apparent linear dependence.

as a function of the inverse reduced atomic mass. The general trend of a larger optic phonon energy with decreasing mass is clearly visible. In ternary compounds such as $Al_xGa_{1-x}As$, where the optic phonon branches of the binary compounds do not overlap, there are two optic phonon modes present, the GaAs-like and the AlAs-like mode. Finally, in contrast to acoustic phonons, which have a large speed at small values of k, the group velocity of optic phonons vanishes near the zone center so that they hardly move. The same happens at the zone boundary for acoustic as well as for optic phonons.

3. Phonon Density of States and Statistics

In order to take into account the quantum mechanical property of lattice vibrations, we will not use classical, but rather quantum mechanical statistics for the acoustic and optic phonons. The thermal occupation of phonon states is determined by the density of states of the phonon modes and by the Bose-Einstein distribution function. The procedure to calculate $n_{ph}(T)$ is very similar to the one used for electrons. However, the density of states for acoustic phonons as well as the distribution function are very different from the one for electrons.

3.1. Density of states

The density of states can be calculated in the same way as for electrons. However, there are some differences between electrons and phonons, which have to be taken into account. For degenerate phonon modes, the degeneracy factor is 3, since there are one longitudinal and two transverse modes for the acoustic and optic regime. The density of states can then be calculated from

$$g(\hbar\omega) = 3 \int \frac{d^3k}{(2\pi)^3}\, \delta[\hbar\omega - \hbar\omega(\underline{k})] \,. \tag{7.28}$$

The dispersion relations for acoustic and optic phonons near $k = 0$ can be expressed according to Eqs. 7.21 and 7.22 as

$$\begin{aligned} \text{acoustic:} \quad & \omega(\underline{k}) = \mathrm{v}_s\, |\underline{k}| \\ \text{optic:} \quad & \omega(\underline{k}) = \omega_0 - \frac{\omega_1}{2}\, \underline{k}^2 \ \text{ for } \hbar\omega < \hbar\omega_0 \end{aligned} \;, \tag{7.29}$$

where $\omega_0 = \omega(k = 0)$. For acoustic phonons, the integration is straightforward, and one obtains near $k = 0$

$$g_{ac}(\hbar\omega) = \frac{3}{2\pi^2\, \hbar}\, \frac{\omega^2}{\mathrm{v}_s^3} \,. \tag{7.30}$$

For the optic branch, the calculation is similar to the one for electrons. The result is

$$g_{op}(\hbar\omega) = \frac{3}{4\,\pi^2} \left(\frac{2}{\hbar\omega_1} \right)^{3/2} \sqrt{\hbar\omega_0 - \hbar\omega} \,. \tag{7.31}$$

The energy dependence of the density of states of optic phonons is therefore similar to the one of electrons at the top of the valence band (cf. paragraph after Eq. 5.9).

3.2. The Bose-Einstein distribution for phonons

In contrast to electrons, phonons have a spin of one. Therefore, the thermal occupation of phonon states follows the Bose-Einstein distribution function $f_{BE}(E,T)$

$$f_{BE}(E,T) = \frac{1}{\exp[(E-\mu)/k_BT] - 1} . \tag{7.32}$$

The chemical potential for phonons vanishes, since the number of particles is not an invariant, i.e., the free energy does not contain the particle number N ($dF/dN = 0 = \mu$). The Bose-Einstein distribution function is shown in Fig. 7.5 as a function of energy for several temperatures assuming a vanishing chemical potential. The behavior for energies much larger than k_BT is the same as for the Fermi-Dirac distribution. However, in the opposite case, when the energy becomes much smaller than the thermal energy, $f_{BE}(E,T)$ exhibits a very different behavior

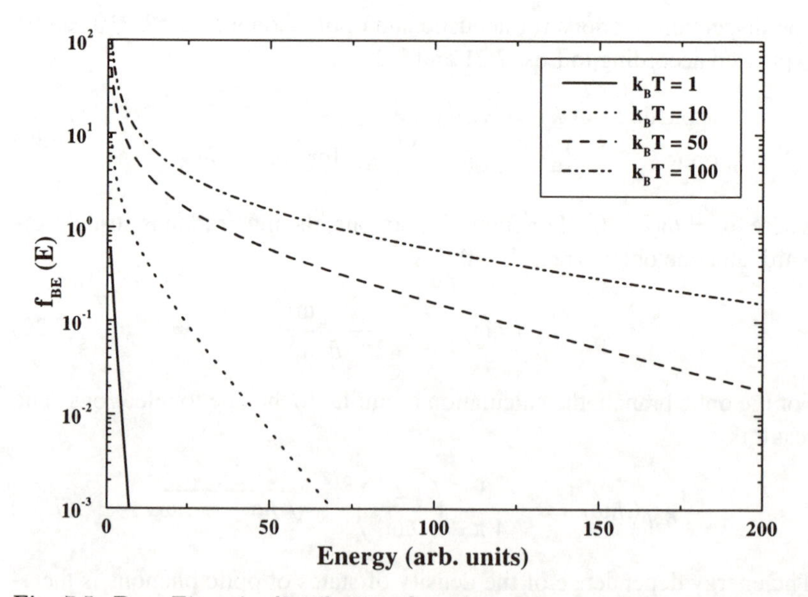

Fig. 7.5. Bose-Einstein distribution function vs energy for several temperatures assuming a vanishing chemical potential.

$$\begin{aligned} E \gg k_B T: \quad & f_{BE} \sim \exp(-\frac{E}{k_B T}) \\ E \ll k_B T: \quad & f_{BE} \sim \frac{k_B T}{E} \end{aligned} \quad . \tag{7.33}$$

The density of thermally excited phonons can be calculated in the same way as the density of electrons (cf. Eq. 5.24)

$$n(T) = \int_0^\infty d(\hbar\omega)\, g(\hbar\omega)\, f_{BE}(\hbar\omega, T)\,. \tag{7.34}$$

Using the acoustic density of states, the density of acoustic phonons $n_{ac}(T)$ can be determined as a function of temperature according to

$$n_{ac}(T) = \frac{3}{2\pi^2\,(\hbar\, v_s)^3} \int_0^\infty \frac{d(\hbar\omega)\,(\hbar\omega)^2}{\exp(\hbar\omega/k_B T) - 1}\,. \tag{7.35}$$

The integration leads to the following result

$$n_{ac}(T) = \frac{3}{2\pi^2}\,\Gamma(3)\,\zeta(3)\left(\frac{k_B T}{\hbar\, v_s}\right)^3 , \tag{7.36}$$

where $\Gamma(3) = 2$ denotes the Gamma function and $\zeta(3) = 1.202$ the Riemann zeta function. The temperature dependence of the acoustic phonon density is therefore a cubic power law. Using a typical sound velocity of $v_s = 5 \times 10^3$ m/s, $n_{ac}(T) = 6.56 \times 10^{15}\ T^3\ \text{cm}^{-3}$, where the temperature is given in Kelvin. The respective temperature dependence between 1 and 330 K is shown in Fig. 7.6 for GaAs and Si. Even at very low temperatures, there is already a significant density of acoustic phonons present.

The same calculation can be performed for optic phonons. Since the frequency or energy range of optic phonons is very limited, a δ-function like distribution can be used to model the phonon density of states. In this case, the optic phonon density $n_{op}(T)$ follows the Bose-Einstein distribution

$$n_{op}(T) = \frac{n_0}{\exp(\hbar\omega_{op}/k_B T) - 1}, \tag{7.37}$$

where n_0 denotes the number of unit cells per volume. The dashed curve in Fig. 7.6 reproduces the phonon density for a material with an optic phonon

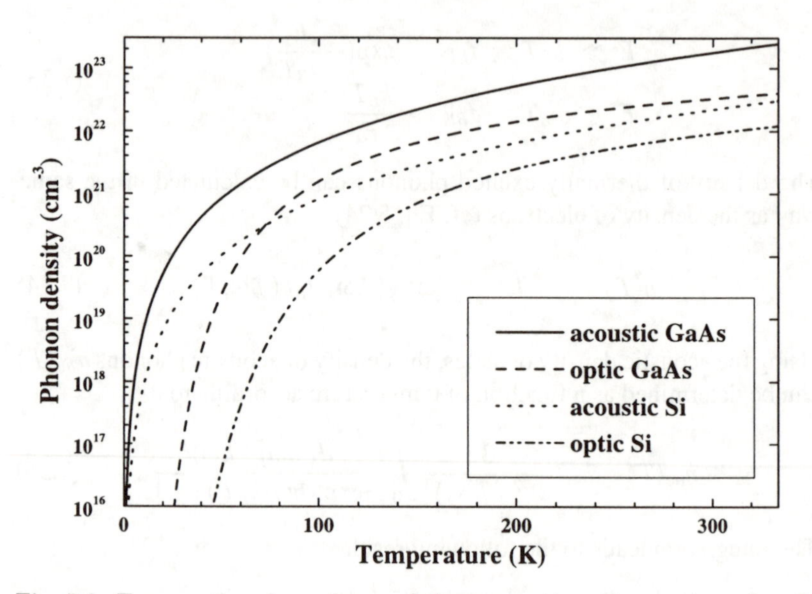

Fig. 7.6. Temperature dependence of the occupation of acoustic and optic phonon states.

energy of 36 meV (e.g., GaAs). In contrast to the acoustic phonons, $n_{op}(T)$ has only a significant value above a certain temperature, which depends on the energy of the optic phonon, i.e., $k_B T > \hbar\omega_{op}$. For Si, the onset temperature for the optic phonon population is larger than for GaAs, since it has a larger optic phonon energy. At low temperatures ($k_B T \ll 36$ meV), the Maxwell-Boltzmann distribution can be used as an approximation.

CHAPTER 8
SCATTERING PROCESSES

In this chapter, the different scattering processes, which determine the temperature dependence of the mobility, will be discussed. The mobility is closely related to the speed of electronic devices such as transistors. It is therefore important to understand the underlying scattering processes. We will consider scattering by acoustic and optic phonons as well as by neutral and ionized impurities. We will then examine the dominating scattering mechanism for the mobility within a particular temperature range for several classes of semiconductors. Finally, we will briefly discuss carrier-carrier scattering and intervalley transfer. For a detailed account of scattering processes in semiconductors, the book by K. Seeger on *Semiconductor Physics* is recommended.

Since the actual calculation of the relaxation time for all these different scattering mechanisms could fill a whole book by itself, we will focus on the actual results and not so much on the calculation. In Chapter 6, the temperature dependence of the relaxation time was given by the following expression in the limit of Maxwell-Boltzmann statistics

$$\langle \tau(T) \rangle = \frac{\int_0^\infty dE\, E^{3/2}\, \tau(E)\, f_0(E)}{\int_0^\infty dE\, E^{3/2}\, f_0(E)} . \tag{8.1}$$

Assuming a power law energy dependence of the scattering time $\tau(E) = \tau_0\, E^n$, the integration results in ($x = E/k_B T$)

$$\langle \tau(T) \rangle = \tau_0\, (k_B T)^n \frac{\int_0^\infty dx\, x^{n+3/2}\, f_0(x)}{\int_0^\infty dx\, x^{3/2}\, f_0(x)} . \tag{8.2}$$

The two integrals in this equation do not contain the temperature explicitly anymore so that the ratio of these integrals results only a numerical factor. For the Maxwell-Boltzmann distribution, the integrals can be directly evaluated leading to

$$\langle\tau(T)\rangle \;=\; \tau_0\,\frac{\Gamma(n+\frac{5}{2})}{\Gamma(\frac{5}{2})}\,(k_BT)^n\,. \tag{8.3}$$

If the energy dependent scattering time is also explicitly temperature dependent, i.e., $\tau = \tau_0^*\,E^n\,(k_BT)^m$, the integration leads to

$$\langle\tau(T)\rangle \;=\; \tau_0^*\,\frac{\Gamma(n+\frac{5}{2})}{\Gamma(\frac{5}{2})}\,(k_BT)^{n+m}\,. \tag{8.4}$$

In the case of the Fermi-Dirac distribution, we have to replace in Eq. 6.22 the numerator f_0 again by the derivative of Eq. 6.21. The result is

$$\langle\tau(T)\rangle \;=\; -\frac{2}{3}\tau_0\,\frac{\int_0^\infty dE\,E^{n+3/2}\,\frac{\partial f_{FD}(E)}{\partial E}}{\int_0^\infty dE\,E^{1/2}\,f_{FD}(E)}\,. \tag{8.5}$$

Since the Fermi-Dirac distribution vanishes for large energies and the power law vanishes for $E = 0$, integration by parts can be used to solve this integral. The result is

$$\begin{aligned}\langle\tau(T)\rangle \;=\;& \frac{2\,\tau_0}{3}\,(k_BT)^n\,(n+\frac{3}{2})\\ &\times\frac{\int_0^\infty dx\,x^{n+1/2}\,f_{FD}(x-\mu/k_BT)}{\int_0^\infty dx\,x^{1/2}\,f_{FD}(x-\mu/k_BT)}\,.\end{aligned} \tag{8.6}$$

Using the definition in Eq. 5.27 and generalizing it to any index, we can rewrite the result of Eq. 8.6 as

$$\langle\tau(T)\rangle \;=\; \tau_0\,(k_BT)^n\,\frac{\Gamma(n+\frac{5}{2})}{\Gamma(\frac{5}{2})}\,\frac{F_{n+1/2}(\mu/k_B\,T)}{F_{1/2}(\mu/k_BT)}\,, \tag{8.7}$$

where the Fermi integrals are defined as

$$F_j(y) \;=\; \frac{1}{\Gamma(j)}\int_0^\infty dx\,x^j\,f_{FD}(x-y)\,. \tag{8.8}$$

Note that in the limit of Maxwell-Boltzmann statistics the Fermi integral F_j is equal to $\exp(\mu/k_BT)$ and the previous result of Eq. 8.3 is recovered. For Fermi statistics, there is an additional temperature dependence of the scattering time through the Fermi integrals. If we again allow for an explicit temperature dependence of the energy scattering time, i.e., $\tau = \tau_0^* E^n (k_B T)^m$, we obtain

$$\langle \tau(T) \rangle = \tau_0^* \, (k_B T)^{n+m} \, \frac{\Gamma(n+\frac{5}{2})}{\Gamma(\frac{5}{2})} \, \frac{F_{n+1/2}(\mu/k_B\,T)}{F_{1/2}(\mu/k_B T)} \; . \tag{8.9}$$

We will now consider phonon scattering as a first example of an important scattering mechanism limiting the mobility within a certain temperature range.

1. Phonon Scattering

In this section, we will consider scattering by thermal acoustic and optic phonons. As described in the previous chapter, acoustic phonons exhibit always a significant density at temperatures of 1 K and above. However, the density of optic phonons becomes only significant above a certain temperature, which depends on the optic phonon energy. For an optic phonon density of 10^{16} cm^{-3}, this temperature is about 30 K for GaAs and 50 K for Si. Optic phonon scattering should therefore become important at higher temperatures for Si than for GaAs. Furthermore, at low temperatures the electrons do not have sufficient energy to emit optic phonons. Therefore, optic phonon scattering can be neglected at low temperatures, and we focus first on acoustic phonon scattering. With regard to scattering processes with phonons, the nonpolar group IV semiconductors behave quite differently from the polar III-V and II-VI semiconductors.

1.1. Acoustic phonons

There are different interaction mechanisms between acoustic phonons and electrons. The most important one is the deformation potential scattering by LA phonons. The acoustic waves modulate the conduction and valence band edges through a change in the lattice constant. The modulation of the band edge energies is usually expressed in terms of an acoustic deformation potential ε_{ac}. Using the relation $\tau = l_{ac}/\mathrm{v}$, where l_{ac} denotes the electron mean free path, it can be shown that the relaxation time has the following temperature and energy dependence

$$\tau(E,T) = \frac{\pi\, \hbar^4\, c_l}{m^{*2}\, \varepsilon_{ac}^2\, k_B T} \sqrt{\frac{m^*}{2\,E}}, \tag{8.10}$$

where c_l denotes the longitudinal elastic constant. Since according to Eq. 8.4 $n = -1/2$ and $m = -1$, the scattering time exhibits a temperature dependence proportional to $T^{-3/2}$

$$\langle \tau(T) \rangle = \frac{2\sqrt{2\pi}}{3} \frac{\hbar^4\, c_l}{m^{*3/2}\, \varepsilon_{ac}^2} (k_B T)^{-3/2}. \tag{8.11}$$

Using a value of 1×10^{11} J/m^3 for c_l and 10 eV for ε_{ac}, one obtains a scattering time of 10 ps at 100 K.

In non-centrosymmetric crystals, a stress can induce a macroscopic electric polarization. In this case, carriers can be scattered by both LA and TA modes via the piezoelectric interaction. This phenomenon is known as the piezo-electric effect. In this case, the energy dependence of the scattering time can be expressed as

$$\tau(E,T) = \frac{2^{3/2}\, \pi\, \hbar^2\, \varepsilon\, \varepsilon_0}{m^{*1/2}\, e^2\, K^2\, k_B\, T} \sqrt{E}, \tag{8.12}$$

where K denotes a dimensionless piezo-electric coupling coefficient. In this case, $n = 1/2$ and $m = -1$. Therefore, the temperature dependence of the mobility is governed by $T^{-1/2}$

$$\langle \tau(T) \rangle = \frac{16\sqrt{2\pi}}{3} \frac{\hbar^2\, \varepsilon\, \varepsilon_0}{m^{*1/2}\, e^2\, K^2} (k_B T)^{-1/2}. \tag{8.13}$$

Using a value of 10^{-3} for K^2 we obtain for GaAs at 100 K a scattering time of 67 ps, which is much larger than the scattering by acoustic deformation potential.

1.2. Optic phonons

The first process to consider is the deformation potential scattering of optic phonons in nonpolar semiconductors. As for the acoustic phonons, the optic phonons also modulate the lattice constant and thus the band edge energies. We will use the Debye temperature as defined by

$$k_B\Theta_D \quad = \quad \hbar\omega_{LO} \tag{8.14}$$

instead of the optic phonon energy. This temperature corresponds to the largest phonon frequency. The optic deformation potential constant D determines the interaction strength between electrons and optic phonons. The momentum relaxation time has now a much more complicated form

$$\begin{aligned} \tau(E,T) \quad &= \quad \frac{2^{1/2}\,\pi\,\hbar^2\,\rho\,k_B\,\Theta_D}{m^{*3/2}\,D^2\,f_{BE}(\Theta_D)} \\ &\times \quad \frac{1}{\sqrt{E+k_B\Theta_D}+e^{\Theta_D/T}\,Re(\sqrt{E-k_B\Theta_D})}\,, \end{aligned} \tag{8.15}$$

where ρ denotes the density of the crystal. Using Maxwell-Boltzmann statistics, the integral in Eq. 8.1 can be written in the following form

$$\langle\tau(T)\rangle \quad = \quad \frac{4\,\sqrt{2\pi}\,\hbar^2\,\rho\,\sqrt{k_B\,\Theta_D}}{3\,m^{*3/2}\,D^2}\,f(\Theta_D/T)\,. \tag{8.16}$$

The function $f(z)$ is given by

$$f(z) = z^{5/2}\,(e^z-1)\int_0^\infty dy \frac{y^{3/2}\,e^{-yz}}{\sqrt{y+1}\,+\,e^z\,Re(\sqrt{y-1})}\,. \tag{8.17}$$

The temperature dependence of the scattering time is mainly determined by the prefactor of the integral, i.e., a good approximation to use is

$$\begin{aligned} \langle\tau(T)\rangle \quad &= \quad \frac{4\,\sqrt{2\pi}\,\hbar^2\,\rho\,\sqrt{k_B\,\Theta_D}}{3\,m^{*3/2}\,D^2} \\ &\times \quad (\frac{\Theta_D}{T})^{5/2}\,(\exp(\Theta_D/T)-1)\,. \end{aligned} \tag{8.18}$$

At low temperature, the exponential function dominates, while for high temperatures the expansion of the exponential function leads a power law with an exponent of

$$\langle\tau(T)\rangle \quad \sim \quad (\frac{\Theta_D}{T})^{7/2}\,. \tag{8.19}$$

Using an optic phonon energy of 36.2 meV, which corresponds to a Debye temperature of 420 K, a density of 5000 kg/m^3, and a deformation potential constant of 5×10^{11} eV/m, one obtains in GaAs at 100 K a value of 360 ps for this scattering time, which is even longer than the previous acoustic phonon scattering times.

In polar semiconductors, the interaction of carriers with optic phonons is known as polar optic phonon scattering. As in the case of piezoelectric scattering, an electric field is induced by the longitudinal optic lattice vibrations. At low temperatures, the energy dependence of the scattering time is given by

$$\tau(E,T) = \frac{1}{2\,\beta\,\omega_{op}} \exp(\Theta_D/T)\,, \tag{8.20}$$

where β denotes the dimensionless constant that measures the coupling of the electrons with the optic phonons

$$\beta = \frac{e\,\hbar\,F_0}{\sqrt{2m^*}\,(\,\hbar\omega_{op})^{3/2}}\,. \tag{8.21}$$

F_0 determines the effective field strength and is given by

$$F_0 = \frac{e\,m^*\,k_B\Theta_D}{4\pi\,\varepsilon_0\,\hbar^2}\,(\varepsilon_{opt}^{-1} - \varepsilon^{-1})\,, \tag{8.22}$$

where ε_{opt} and ε denote the high and low frequency dielectric constants, respectively. Using the energy dependence of Eq. 8.20, we obtain an average relaxation time in the limit $T \ll \Theta_D$ of

$$\langle\tau(T)\rangle = \frac{1}{2\,\beta\,\omega_{op}} \exp(\Theta_D/T)\,. \tag{8.23}$$

For GaAs, the dimensionless polar coupling constant has a value of 0.067 resulting in a scattering time of 57 ps at 100 K. The general expression of the energy dependence of the scattering time in this case is rather lengthy. Using the definitions $a = (1 + \hbar\omega_{op}/E)^{1/2}$ and $b = Re(1 - \hbar\omega_{op}/E)^{1/2}$, the expression for $\tau(E,T)$ is

$$\begin{aligned}\tau(E,T) &= \frac{(e^{\Theta_D/T} - 1)}{\beta\,\omega_{op}}\,(\frac{E}{\hbar\omega_{op}})^{1/2} \\ &\times \left[\ln\left|\frac{a+1}{a-1}\right| + e^{\Theta_D/T}\ln\left|\frac{1+b}{1-b}\right|\right]^{-1}.\end{aligned} \tag{8.24}$$

For high temperatures ($T \gg \Theta_D$), an approximation derived from the numerical solution of the integral for the scattering time is given by

$$\langle \tau(T) \rangle \;=\; \frac{4}{3\sqrt{\pi}\,\beta\,\omega_{op}} \frac{T}{\Theta_D} \exp(\Theta_D/T)\,. \tag{8.25}$$

For an unrealistic temperature of 2000 K, which fulfills the above condition, we obtain a scattering time of 7.5 ps.

2. Impurity Scattering

Electron and holes are also scattered by impurities, which can be neutral or charged. In doped materials, carrier scattering by ionized impurities will limit the scattering time or mobility. However, at very low temperatures, all impurities will become neutral due to carrier freeze-out. Since the carrier relaxation time due to phonon scattering increases with decreasing temperature, the residual impurities will eventually determine the scattering time. We will first consider the contribution of neutral impurities and then briefly discuss the effect of ionized impurities.

2.1. Neutral impurities

In the case of neutral impurities, the carrier is deflected by short-range interaction with the impurity atom itself, and not by the long-range Coulomb interaction with its charge. The numerical, quantum mechanical calculation of the scattering cross section includes exchange effects as well as the effect of the polarization of the atom by the incident carrier. The relaxation time has the following form

$$\tau(E,T) \;=\; \frac{e^2}{80\,\pi\,\varepsilon_0\,\hbar^3} \frac{m^{*2}}{\varepsilon\,N_I}\,, \tag{8.26}$$

where N_I denotes the density of neutral impurities. Note that τ neither depends on energy, nor on temperature so that the resulting average scattering time is temperature independent with the value given in Eq. 8.26. Since this scattering time is independent of temperature, it will limit the mobility at very low temperatures. A typical value of τ for impurity scattering for a density of $N_I = 10^{16}$ cm^{-3} in GaAs is 0.3 ps, which is much shorter than the phonon scattering times.

2.2. *Ionized impurities*

In the case of ionized impurity scattering, the carrier and the charged impurity interact directly via the Coulomb interaction, leading to a scattering time of the form

$$\tau(E,T) \quad = \quad \frac{16\pi\,\sqrt{2m^*}}{N_I}\left(\frac{\varepsilon\varepsilon_0}{Z\,e^2}\right)^2 E^{3/2}\,, \qquad (8.27)$$

where Ze denotes the charge of the impurity ion. This expression has to be multiplied with a correction factor, which is weakly energy dependent. Therefore, the temperature dependence of the scattering time will be dominated by a power law exponent of 3/2. Within this approximation, the expression for the average scattering time is

$$\langle\tau(T)\rangle \quad = \quad \frac{128\,\sqrt{2\pi}\,m^*\,(\varepsilon\varepsilon_0)^2}{Z^2\,e^4\,N_I}\,(k_BT)^{3/2}\,. \qquad (8.28)$$

At a temperature of 100 K using the parameters of GaAs, we obtain a value of 7 ps, which is longer than the neutral impurity scattering time.

3. Temperature Dependence of the Mobility

The mobility is related to the average scattering time through

$$\mu(T) \quad = \quad \frac{e}{m^*}\,\langle\tau(T)\rangle\,. \qquad (8.29)$$

For GaAs a scattering time of 10 ps results in a mobility of 26.3 T^{-1}, which corresponds to 263,000 cm^2/Vs. In Tab. 8.1 the different scattering processes and their respective temperature dependencies are listed. A typical scattering time and its corresponding mobility for the effective mass of GaAs have been added. In Figs. 8.1 and 8.2 the temperature dependence of the scattering time is shown for non-polar and polar semiconductors, respectively. The time scale is arbitrary. These figures are only used to demonstrate the very different temperature dependence of these scattering

Table 8.1. Temperature dependence of the mobility for different scattering mechanisms. The temperature dependence together with the corresponding prefactor are listed. The scattering time at 100 K for the parameters of GaAs is also included.

Scattering process	Prefactor	Temperature dependence	τ(100 K)
acoust. deform.	$\frac{2\sqrt{2\pi}\, e\, \hbar^4\, c_l}{3\, m^{*5/2}\, \varepsilon_{ac}^2}$	$(k_B T)^{-3/2}$	10 ps
acoust. piezoel.	$\frac{16\sqrt{2\pi}\ \hbar^2\, \varepsilon\varepsilon_0}{3\, m^{*3/2}\, e\, K^2}$	$(k_B T)^{-1/2}$	67 ps
optic deform.	$\frac{4\sqrt{2\pi}\ \hbar^2\, e\, \rho\, \sqrt{k_B \Theta_D}}{3\, m^{*5/2}\, D^2}$	$(\frac{\Theta_D}{T})^{5/2} \times (e^{\Theta_D/T} - 1)$	360 ps
polar optic	for $T \ll \Theta_D$: $\frac{1}{2\, \beta\, \omega_{op}}$	$\exp(\Theta_D/T)$	57 ps
neutral impurity	$\frac{e^3\, m^*}{80\pi\ \hbar^3\, \varepsilon\varepsilon_0\, N_I}$	1	0.3 ps
ionized impurity	$\frac{128\sqrt{2\pi}\, (\varepsilon\varepsilon_0)^2}{Z^2\, \sqrt{m^*}\, e^3\, N_I}$	$(k_B T)^{3/2}$	7 ps

processes. In order to determine the scattering time for a combination of for example acoustic deformation potential and ionized impurity scattering within the relaxation time approximation, first the inverse energy dependent relaxation times have to be added, which implies that the scattering probabilities have to be summed. After this process, the averaging procedure described in the beginning of this chapter can be used. At low temperature, ionized impurity scattering dominates, while at room temperature the scattering time is limited by scattering with optic phonons. In between, scattering by acoustic phonons may prevail.

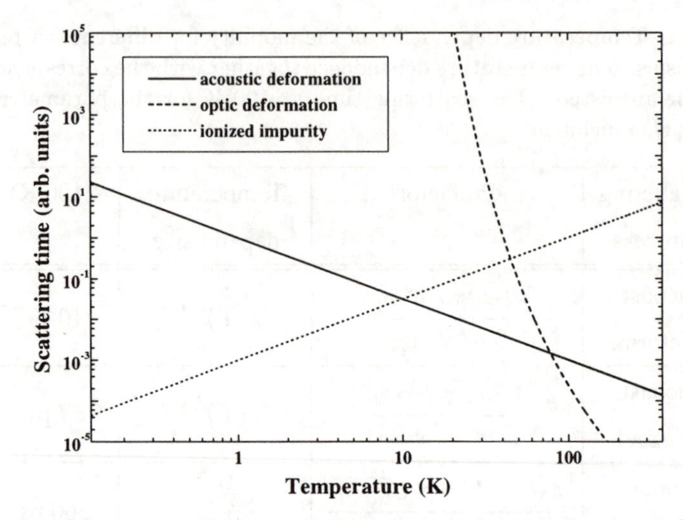

Fig. 8.1. Schematic temperature dependence of scattering times for non-polar semiconductors such as Si and Ge.

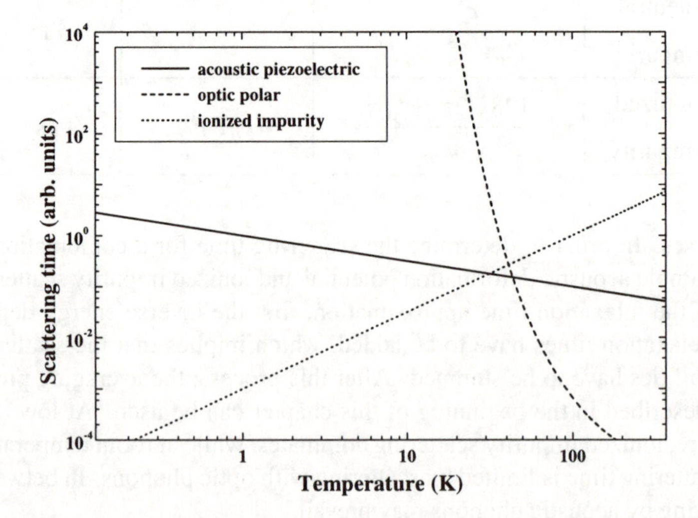

Fig. 8.2. Schematic temperature dependence of scattering times for polar semiconductors such as III-V and II-VI materials.

4. Carrier-Carrier Scattering

We will briefly mentioned carrier-carrier scattering. In a process, where a carrier is scattered by another carrier, the total momentum of the carrier gas is not changed. Therefore, carrier-carrier scattering alone has no significant influence on the mobility. However, in combination with another scattering processes, it sometimes has an important influence. For example, the combination of ionized impurity and carrier-carrier scattering in a non-degenerate semiconductor leads to a reduction of the mobility by a factor of 1.67. This effect depends naturally on the carrier density. The changes for scattering by phonons are less pronounced. In the limit of large carrier densities, the effects of carrier-carrier scattering on the mobility are less important.

5. Intervalley Scattering

So far we have only considered intravalley scattering, i.e., the energy and momentum of the electron were only changed by a small amount. If many valleys exist with the same energy such as in Si and Ge, the carrier energy is not changed by much, only the momentum has to change by a large amount. In a direct gap semiconductor such as GaAs, there is only one minimum at $k = 0$. However, other minima exist at higher energies, e.g., at X and L. If the carriers have sufficient energy, they can be scattered from the Γ-valley to the X- or L-valleys. Since all valley transfer processes require a large change in the momentum of the carrier, it is mediated either by an impurity or by a zone-boundary phonon. Since impurity scattering is only important at low temperatures, the most important mechanisms for intervalley scattering are acoustic and optic phonon scattering.

Scattering between equivalent valleys in nonpolar semiconductors can be treated in a similar way as acoustic and optic phonon deformation potential scattering. The energy of the involved optic phonon is reduced, since a zone boundary phonon is needed. There are actually two different possibilities for scattering between equivalent valleys. The so-called *g process* involves two valleys along the same direction on the opposite side of the Brillouin zone (e.g., from [100] to [$\bar{1}$00]). When a carrier is scattered into a valley in a different direction in reciprocal space (e.g., from [100] to [010]), it is denoted *f process*. In doped Si, the f-type scattering is weak.

The combined relaxation time in Si has been shown to decrease as a power law with an exponent of

$$\langle \tau(T) \rangle \propto (T/\Theta_D)^{-5/2} . \tag{8.30}$$

Applying a large electric field to a semiconductor can result in a redistribution of the carriers in different valleys. This effect is of particular importance in some III-V semiconductors, where the lowest conduction band minimum is at Γ. By applying an electric field, these carriers can acquire enough energy so that they can be transferred into a different minimum, e.g., at the X- or L-point of the Brillouin zone. The transfer probability increases with increasing field reducing the conductivity or drift velocity. A simple analytic expression for this type of drift velocity is

$$\mathrm{v}_d(F_E) = \frac{\mu F_E}{1 + (F_E/F_0)^2}, \tag{8.31}$$

where F_0 denotes a critical field determining the field strength of the maximum of the drift velocity. The value at the maximum is $\mu F_0/2$. This drift velocity-field characteristics is shown in Fig. 8.3. At low fields the drift velocity is proportional to the electric field strength, the drift mobility regime. After a maximum at F_0, the drift mobility decreases again. The drift velocity or mobility of the X- or L-valley is usually smaller than that of the Γ-valley, since the effective mass in these valleys is larger. The dashed-dotted curve shows the drift velocity for the other valleys. The resulting combined drift velocity has a regime, where the drift velocity decreases with increasing field, the negative differential velocity (NDV), negative differential conductivity (NDC) or negative differential resistance regime (NDR). This type of NDV leads to very peculiar effects such as propagating electric-field domains in these types of semiconductors, which result in oscillations of the current. This is known as the Gunn effect, e. g. in GaAs. In addition to this voltage-controlled or N-shape NDR, there exists also current-controlled or S-shape NDR resulting in the formation of current filaments. This effect has been observed, e.g., in n-type Ge during impact ionization.

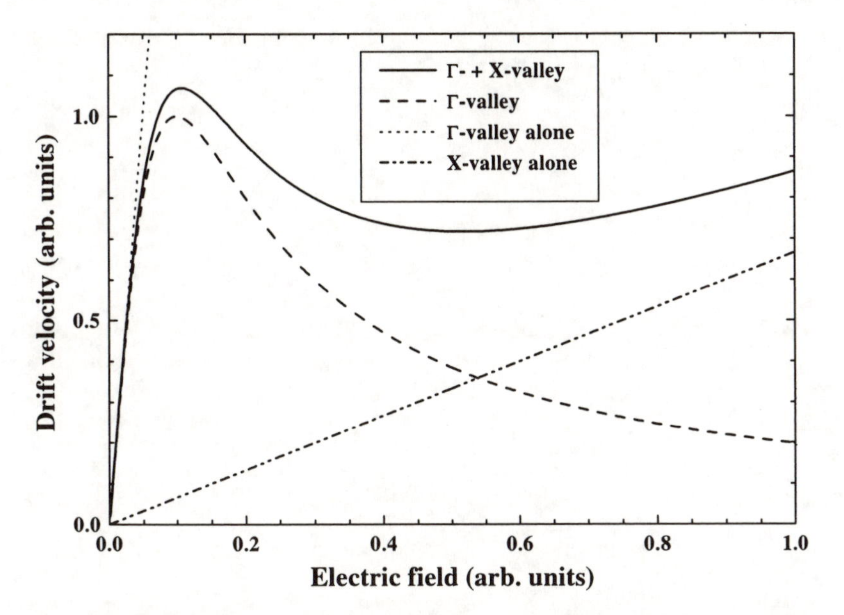

Fig. 8.3. Drift velocity vs electric field for intervalley transfer in some III-V semiconductors.

CHAPTER 9

EXCITONS

Optical excitation creates the same number of electrons in the conduction band as holes in the valence band. Due to the Coulomb attraction between electrons and holes, the energy necessary to create an electron-hole pair is slightly smaller than the energy gap. The difference between the energy gap and the electron-hole pair transition can vary considerably between different materials. Neglecting both exchange and correlation terms, the interaction takes place only through the Coulomb interaction. The resulting correlated electron-hole pair is known as an exciton.

There are two limiting cases. In the case of a very strong electron-hole attraction such as in ionic crystals, the electron and hole are tightly bound to each other within the same or nearest-neighbor unit cell. These excitons are known as Frenkel excitons. In most semiconductors, electrons and holes are only weakly bound to each other. This type of exciton is usually referred to as the Wannier-Mott exciton.

1. Exciton States

Let us consider an electron-hole pair in a semiconductor. In the effective mass approximation, the electron and hole are described as free carriers with effective masses m_e^* and m_h^*, respectively. The interaction takes place via the Coulomb attraction. The Schrödinger equation for this problem can be written in terms of the electron and hole coordinates $\underline{r}_e$ and $\underline{r}_h$, respectively, as

$$\left(-\frac{\hbar^2}{2m_e^*}\Delta_{\underline{r}_e} - \frac{\hbar^2}{2m_h^*}\Delta_{\underline{r}_h} - \frac{e^2}{4\pi\varepsilon\varepsilon_0|\underline{r}_e - \underline{r}_h|}\right)\Psi(\underline{r}_e,\underline{r}_h) = E\,\Psi(\underline{r}_e,\underline{r}_h) \quad , \qquad (9.1)$$

where ε and ε_0 denote the dielectric constant of the material and the vacuum, respectively. Transforming the coordinates into the center-of-mass system with the relative coordinate $\underline{r} = \underline{r}_e - \underline{r}_h$, the reduced effective mass $m_r^* = m_e^* m_h^* / (m_e^* + m_h^*)$, and the center-of-mass coordinate $\underline{R} = (m_e^* r_e + m_h^* r_h)/M^*$ with $M^* = m_e^* + m_h^*$, Eq. 9.1 is transformed into

$$\left(-\frac{\hbar^2}{2m_r^*}\Delta_{\underline{r}} - \frac{\hbar^2}{2M^*}\Delta_{\underline{R}} - \frac{e^2}{4\pi\varepsilon\varepsilon_0 |\underline{r}|} \right) \Psi(\underline{r},\underline{R}) = E\,\Psi(\underline{r},\underline{R}) \,. \quad (9.2)$$

The equation is separable, i.e., the eigenfunctions can be written as

$$\Psi(\underline{r},\underline{R}) = \exp(i\underline{K}\cdot\underline{R})\,\Psi(\underline{r}) \,, \quad (9.3)$$

where $\underline{K}$ denotes the center-of-mass momentum. The center-of-mass motion is a plane wave, and the equation for the relative motion can be written as

$$\left(-\frac{\hbar^2}{2m_r^*}\Delta_{\underline{r}} - \frac{e^2}{4\pi\varepsilon\varepsilon_0 |\underline{r}|} \right) \Psi(\underline{r}) = \left(E - \frac{\hbar^2 \underline{K}^2}{2M^*} \right) \Psi(\underline{r}) \,. \quad (9.4)$$

The solutions of this equation are the well-known wave functions of the hydrogen atom with a reduced effective mass m_r^* and an effective electron charge $e/\sqrt{\varepsilon}$. Taking the energy gap as the reference energy, the exciton energies are therefore given by

$$E_n(\underline{K}) = E_G + \frac{\hbar^2 \underline{K}^2}{2\,M^*} - \frac{R^*}{n^2} \,, \quad (9.5)$$

where R^* denotes the effective Rydberg constant of the exciton

$$R^* = \frac{m_r^*\, e^4}{2(4\pi\varepsilon_0\, \varepsilon\, \hbar)^2} \,. \quad (9.6)$$

Using the binding energy R of the hydrogen atom of 13.6 eV, the Rydberg constant for an exciton can be written as

$$R^* = \frac{m_r^*}{m} \frac{1}{\varepsilon^2} R \,. \quad (9.7)$$

Since m_r^* is typically of the order of $0.1m$ and $1/\varepsilon^2$ of the order of 0.01, a typical exciton energy R^* in a semiconductor is of the order of 10 meV. The

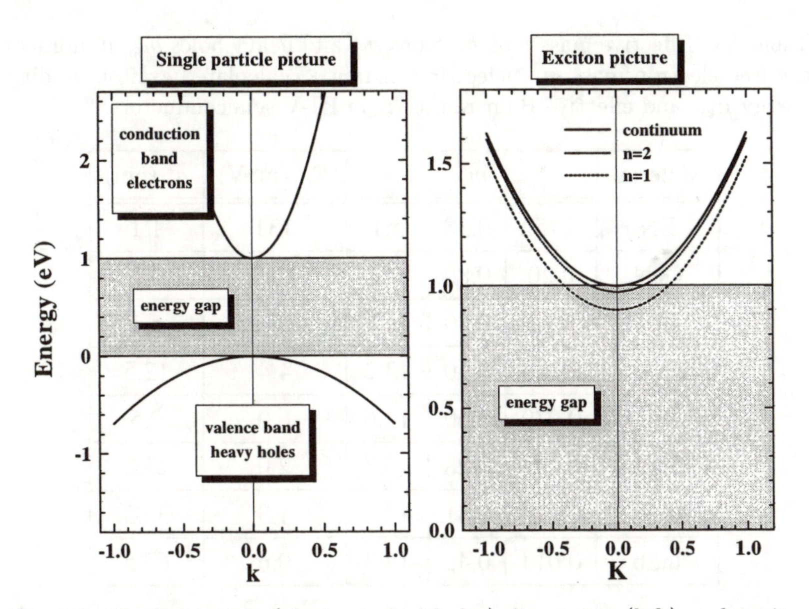

Fig. 9.1. Single particle (electrons and holes) dispersion (left) and exciton dispersion (right) in a direct gap semiconductor. The wavevector and the energy scales are different. The effective mass of the exciton dispersion is given by the total mass M^*.

exciton dispersion for the ground state and the first excited state is shown in comparison to the single particle dispersion of electrons and holes in Fig. 9.1. Note that the exciton dispersion cannot be represented within a single particle picture. This is true not only for the bound states, but also for the continuum states. The exciton states should therefore never be plotted in an energy band diagram containing valence and conduction band.

2. Exciton Binding Energies

The exciton binding energy $E_{ex} = R^*$ is derived from the energy of the exciton ground state $n = 1$. For GaAs, an exciton binding energy of 4.4 meV is obtained. In Tab. 9.1 the effective electron and hole masses as well as the dielectric constants are compiled for a number of III-V semiconductors. The larger the mass, the larger the binding energy. The di-

Table 9.1. Effective masses of electrons m_e^* and heavy holes m_{hh}^* in units of the free electron mass m, dielectric constant ε, calculated exciton binding energy E_{ex}, and effective Bohr radius a^* in III-V semiconductors.

Material	m_e^*	m_{hh}^*	ε	E_{ex} (meV)	a^* (nm)
BN	0.752	0.38	5.1	131	1.1
GaN	0.20	0.80	9.3	25.2	3.1
InN	0.12	0.50	9.3	15.2	5.1
GaAs	0.063	0.50	13.2	4.4	12.5
InP	0.079	0.60	12.6	6.0	9.5
GaSb	0.041	0.28	15.7	2.0	23.2
InAs	0.024	0.41	15.2	1.3	35.5
InSb	0.014	0.42	17.3	0.6	67.5

electric constant enters quadratically into the exciton binding energy and therefore can also have a significant effect on it. For III-V semiconductors, typical values are several meV, while GaN and InN have a binding energy above 10 meV. The effective mass of light holes is smaller than the heavy-hole mass. Therefore, the exciton binding energy formed by electrons and light holes is smaller than the one for heavy-hole excitons. Some wide gap semiconductors have exciton binding energies of several tens meV. In particular II-VI (Tab. 9.2) and I-VII semiconductors (Tab. 9.3) exhibit rather large exciton binding energies. In some materials, the binding energy exceeds the thermal energy at room temperature so that excitons can even be observed at room temperature. In some I-VII semiconductors, the binding energy can even reach 100 meV. Note that, although these are rather large energies for excitons, they are still about 1.5 orders of magnitude smaller than the energy gap.

The spatial extent of the exciton can be estimated from its Bohr radius. Using again the result for the hydrogen atom, the Bohr radius is given by

$$a_n^* = \frac{4\pi\,\varepsilon_0\,\varepsilon\,\hbar^2}{m_r^*\,e^2}\,n = \frac{\varepsilon}{m_r^*/m}\,a_B\,n\,, \tag{9.8}$$

Table 9.2. Effective masses of electrons m_e^* and heavy holes m_{hh}^* in units of the free electron mass m, dielectric constant ε, calculated exciton binding energy E_{ex}, and effective Bohr radius a^* in II-VI semiconductors.

Material	m_e^*	m_{hh}^*	ε	E_{ex} (meV)	a^* (nm)
ZnS (c)	0.34	1.76	8.9	49.0	1.7
ZnO	0.28	0.59	7.8	42.5	2.2
ZnSe (c)	0.16	0.78	7.1	35.9	2.8
CdS (h)	0.21	0.68	9.4	24.7	3.1
ZnTe	0.12	0.6	8.7	18.0	4.6
CdSe (h)	0.11	0.45	10.2	11.6	6.1
CdTe	0.096	0.63	10.2	10.9	6.5
HgTe	0.031	0.32	21.0	0.87	39.3

where $a_B = 5.29 \times 10^{-11}$ m denotes the Bohr radius of the hydrogen atom. For GaAs, the Bohr radius a^* of the ground state exciton ($n = 1$) has a value of 12.5 nm, which is much larger than the unit cell in this material. The larger the effective mass, the smaller the Bohr radius. For I-VII semiconductors, the effective Bohr radius is still a few times the atomic separation so that the excitons in these materials are considered to be Wannier-Mott excitons. In all III-V and II-VI semiconductors, the effective Bohr radius is

Table 9.3. Effective masses of electrons m_e^* and heavy holes m_{hh}^* in units of the free electron mass m, dielectric constant ε, calculated exciton binding energy E_{ex}, and effective Bohr radius a^* in I-VII semiconductors.

Material	m_e^*	m_{hh}^*	ε	E_{ex} (meV)	a^* (nm)
CuCl	0.43	4.2	7.9	85.1	1.1
CuI	0.33	1.4	6.5	86.1	1.3
CuBr	0.28	1.4	7.9	50.9	1.8

between one and two orders of magnitude larger than the atomic separation. The Bohr radii of the ground state excitons are also shown in Tabs. 9.1, 9.2, and 9.3 for III-V, II-VI, and I-VII semiconductors, respectively.

In Fig. 9.2 the exciton binding energies for a number of III-V, II-VI, and I-VII semiconductors are plotted versus the energy gap. A general trend of an increasing binding energy with increasing gap energy can be observed. According to the $\underline{k} \cdot \underline{p}$ theory, a larger gap is correlated with a larger effective mass, which in turn results in a larger exciton binding energy. The dependence of the exciton Bohr radius on the energy gap is shown in Fig. 9.3 for direct-gap III-V, II-VI, and I-VII semiconductors. The Bohr radius decreases with increasing energy gap, i.e., with increasing effective mass.

We have used the zero frequency dielectric constant to calculate the exciton binding energy. This approximation is only valid as long as the binding energy is significantly smaller than the longitudinal optical (LO) phonon energy in the corresponding material. In particular, this approxi-

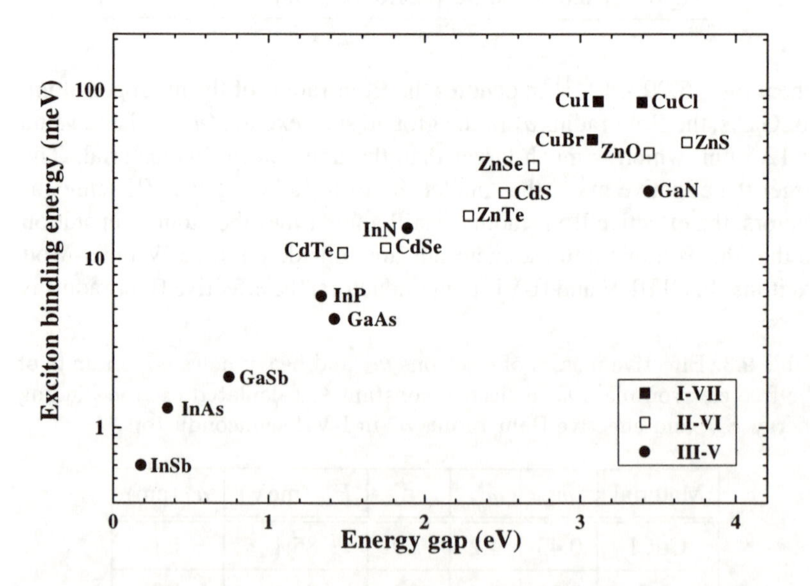

Fig. 9.2. The exciton binding energy as a function of the energy gap in several III-V, II-VI, and I-VII semiconductors.

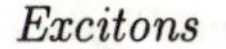

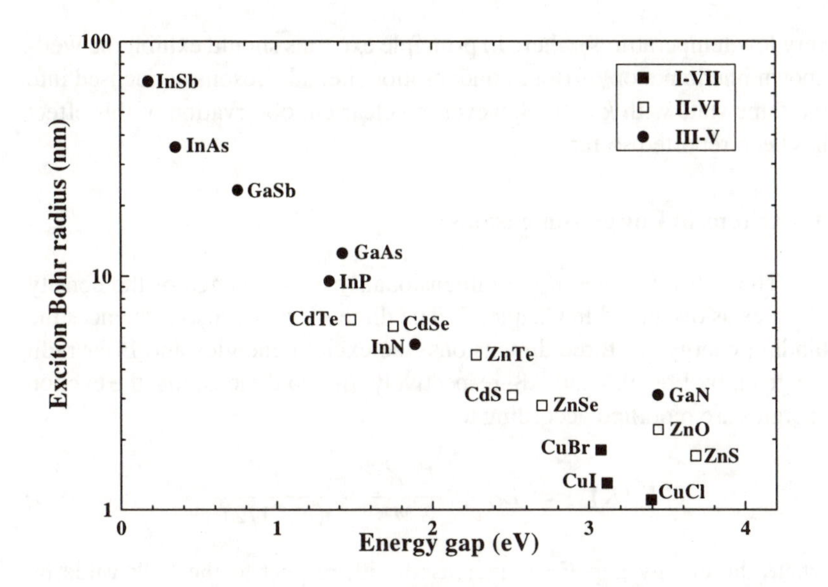

Fig. 9.3. The exciton Bohr radius as a function of the energy gap in several III-V, II-VI, and I-VII semiconductors.

mation may not be valid for the excitonic ground state in wide energy-gap materials. The dielectric constant at high frequencies is usually smaller than the static dielectric constant. Therefore, the use of the static dielectric constant leads to an underestimation of the true binding energy.

Indirect semiconductors also exhibit exciton states. The determination of exciton binding energies is, however, not as simple as for direct semiconductors, because the wavevector of the electron and hole are different. As a first approximation, the exciton dispersion follows the dispersion of the energy gap.

The combination of an electron-hole pair in an exciton leads, in the picture of second quantization, to the formation of a bosonic-type particle. This has two consequences. The first one occurs for low densities and not too low temperatures, where the excitons can be described by Maxwell-Boltzmann statistics. With increasing density, the deviation from ideal Bosons increases until at very high densities an electron-hole plasma made up entirely of Fermions is formed. The second consequence appears at very low temperatures, where in principle excitons should exhibit the well-known phenomenon of Bose condensation, i.e., all Bosons condensed into the same state with $\underline{k} = \underline{0}$. However, no clear-cut observation of this effect has been reported so far.

3. Excitons in Lower Dimensions

The first influence of the dimensionality is the change of the density of states as discussed in Chapter 5. The dimensionality also influences the binding energy. In three dimensions, the exciton energies and Bohr radii are given by Eqs. 9.5 and 9.8, respectively. In two dimensions, the exciton energies are modified according to

$$E_n(\underline{K}) = E_G + \frac{\hbar^2 \underline{K}^2}{2\, M^*} - \frac{R^*}{(n-1/2)^2}, \tag{9.9}$$

where the energy gap E_G is increased with respect to the bulk value by the confinement energy E_n^{co}. When the dimension is reduced from three to two, the binding energy of the ground state is therefore enhanced, i.e., $E_{ex}^{2D} = 4E_{ex}^{3D}$. The effective Bohr radius in two dimensions becomes

$$a^* = \frac{\varepsilon}{m_r^*/m}\, a_B \left(n - \frac{1}{2}\right) \tag{9.10}$$

so that the Bohr radius for the ground state is reduced by a factor of two, i.e., $a_{2D}^* = a_{3D}^*/2$. Note that in quasi-2D systems the degeneracy of the heavy- and light-hole states at the Γ-point in cubic semiconductors is removed. At the same time, the in-plane dispersion of heavy- and light-hole states can be reversed so that the ground state, which originates from the heavy-hole bulk state, acquires a light-hole mass, while the light-hole state becomes heavy.

In strictly one and zero dimensions, the exciton binding energy diverges. Therefore, it is not possible to give explicit formulas for the exciton energies. However, real quantum wires and dots are always three-dimensional so that the divergence cannot be observed. In order to determine the exciton binding energy for real quantum wires and dots, numerical calculations have to be performed.

CHAPTER 10

OPTICAL ABSORPTION AND EMISSION

Many applications of semiconductors in optical devices are associated with the fact that they exhibit energy gaps from the far-infrared into the visible wavelength regime. The two most prominent ones are light detectors and light emitters. While Si completely dominates the electronic applications, it is much less important for optical applications, since it exhibits an indirect energy gap. The direct energy-gap materials consisting of III-V and II-VI semiconductors are far more useful for optical applications. In this chapter, we will first discuss the absorption properties of semiconductors excluding excitonic effects. Then, we will summarize the influence of excitons on the absorption spectrum and present the absorption coefficient for semiconductors structures with dimensions lower than three. Finally, we will review the emission properties of semiconductors including luminescence spectra and radiative lifetimes.

1. Absorption without Excitons

In order to determine the absorption properties, we have to determine the response of the semiconductor to electromagnetic radiation. The Hamiltonian for an electron in a solid interacting with an electromagnetic wave is given by

$$H(\underline{p},\underline{r},\underline{A}) \quad = \quad \frac{1}{2\,m}\left(\underline{p} + e\,\underline{A}\right)^2 + V(\underline{r})\,, \tag{10.1}$$

where $\underline{A}(\underline{r},t)$ denotes the vector potential. Choosing the Coulomb gauge, the electrostatic potential $\phi = 0$ and $\underline{\nabla} \cdot \underline{A} = 0$. The electric field strength $\underline{F}_E$ and magnetic field strength $\underline{B}$ of the light wave are given by

$$\underline{F}_E = -\frac{\partial \underline{A}}{\partial t} \quad \text{and} \quad \underline{B} = \underline{\nabla} \times \underline{A}\,, \tag{10.2}$$

respectively. The momentum operator does not commute with the vector potential. However, since the gauge was chosen such that $\underline{\nabla} \cdot \underline{A} = 0$, the quadratic term in Eq. 10.1 leads to

$$H(\underline{p},\underline{r},\underline{A}) = H_0(\underline{p},\underline{r}) + \frac{e}{m}\underline{A} \cdot \underline{p} + \frac{e^2}{2\,m}\underline{A}^2 \,. \qquad (10.3)$$

Since we are only interested in the linear response of the system, i.e., the linear optical properties, we can neglect in first order perturbation theory the quadratic term in Eq. 10.3. The resulting perturbation consisting of the second term in Eq. 10.3 is sometimes referred to as the electron-radiation interaction Hamiltonian H_{er}

$$H_{er}(\underline{p},\underline{A}) = \frac{e}{m}\underline{A} \cdot \underline{p} \,. \qquad (10.4)$$

Note that the actual form of the operator depends on the gauge we are using. The actual physical result, however, will be independent of the gauge. Assuming a plane wave for the light, the vector potential and electric field are related through

$$\underline{A}(\underline{r},t) = -\frac{\underline{\hat{e}}\,F_E}{2\,\omega}\left\{\exp[i(\underline{q} \cdot \underline{r} - \omega t)] + \text{c.c.}\right\} , \qquad (10.5)$$

where $\underline{\hat{e}}$ denotes the polarization vector, ω the frequency of the light wave, $\underline{q}$ its wavevector, and c.c. the complex conjugate. Using Fermi's Golden Rule as the standard approach to time-dependent perturbation theory, the quantity, which has to be determined, is the transition probability R for absorption given by

$$R(\hbar\omega) = \frac{2\pi}{\hbar}\sum_{\underline{k}_C,\underline{k}_V} |\langle C|H_{er}|V\rangle|^2 \; [\delta(E_C(\underline{k}_C) - E_V(\underline{k}_V) - \hbar\omega)$$
$$+ \; \delta(E_C(\underline{k}_C) - E_V(\underline{k}_V) + \hbar\omega)] \,. \qquad (10.6)$$

The δ-functions ensure energy conservation. The term with the minus sign in front of the photon energy corresponds to the absorption signal, the term with the plus sign describes stimulated emission, i.e., emission in the presence of an external field. In the following, we will only consider the absorption term. The photon carries a very small momentum $\underline{q}$ compared to a typical wavevector of the first Brillouin zone of the solid. Optical transitions are therefore almost vertical in $\underline{k}$-space, i.e., the initial and final value for $\underline{k}$ are almost the same. It can be shown that this condition is actually exact for electric dipole transitions, which corresponds to the approximation $\exp[i(\underline{q}\cdot\underline{r})] = 1$. In this case $\underline{k}_V$=$\underline{k}_C$, and the transitions are exactly vertical or direct. For this type of transitions, the expectation value of the perturbation operator can be written as

$$|\langle C|H_{er}|V\rangle|^2 \;=\; \left(\frac{e}{m\,\omega}\right)^2 \left|\frac{F_E}{2}\right|^2 |P_{CV}|^2 \,, \qquad (10.7)$$

where the momentum matrix element $P_{CV} = \langle C|\underline{\hat{e}}\cdot\underline{p}|V\rangle$ was assumed to be independent of $\underline{k}$. Within Kane's model of the energy band structure, $|P_{CV}|^2 = m^2P^2/(3\,\hbar^2)$. The transition rate per unit volume is then given by

$$R(\hbar\omega) \;=\; \frac{2\pi}{\hbar}\left(\frac{e}{m\,\omega}\right)^2 \left|\frac{F_E}{2}\right|^2 |P_{CV}|^2 \times \sum_{\underline{k}} \delta(E_C(\underline{k}) - E_V(\underline{k}) - \hbar\omega)\,. \qquad (10.8)$$

This transition rate per unit volume multiplied by the photon energy $\hbar\omega$ is simply the power loss of the incident light due to absorption in the unit volume. The power loss is directly connected with the absorption coefficient α. The incoming light intensity I is related to the electric field by

$$I \;=\; \frac{\varepsilon_0}{2}\,c\,n_r\,|F_E|^2\,, \qquad (10.9)$$

where n_r denotes the index of refraction. The power loss $R\times\hbar\omega$ is given by the change in intensity

$$\frac{c}{n_r}R\times\hbar\omega = -\frac{dI}{dt} \;=\; -\left(\frac{dI}{dx}\right)\left(\frac{dx}{dt}\right) \;=\; \frac{c}{n_r}\alpha\,I\,. \qquad (10.10)$$

The absorption coefficient can therefore be expressed as

$$\alpha(\hbar\omega) \;=\; \frac{\hbar\omega}{I}R(\hbar\omega)\,. \qquad (10.11)$$

Using Eqs. 10.8 and 10.9, we obtain for the absorption coefficient

$$\alpha(\hbar\omega) = \frac{\pi e^2}{\varepsilon_0 m^2 c n_r \omega} |P_{CV}|^2 \times \sum_{\underline{k}} \delta(E_C(\underline{k}) - E_V(\underline{k}) - \hbar\omega) . \qquad (10.12)$$

The coefficient in front of the sum

$$\frac{\pi e^2}{\varepsilon_0 m^2 c n_r \omega} |P_{CV}|^2 \qquad (10.13)$$

can be rewritten in terms of the oscillator strength f_{VC} of the optical transition, which is determined by the momentum matrix element, i.e.,

$$f_{VC}(\hbar\omega) = \frac{2 |P_{CV}|^2}{m \hbar\omega} . \qquad (10.14)$$

Within Kane's model presented in Chapter 4, the oscillator strength can be expressed using Eq. 4.29 as

$$f_{VC}(\hbar\omega) = \frac{\frac{m}{m_e^*} - 1}{3 \hbar\omega} \frac{E_G + \Delta}{1 + \frac{2 \Delta}{3 E_G}} . \qquad (10.15)$$

With this dimensionless quantity the absorption coefficient becomes

$$\alpha(\hbar\omega) = C \frac{f_{VC}}{n_r} \sum_{\underline{k}} \delta(E_C(\underline{k}) - E_V(\underline{k}) - \hbar\omega) , \qquad (10.16)$$

where $C = \frac{\pi e^2 \hbar}{2 \varepsilon_0 m c}$ is independent of any material parameters. If the electric dipole matrix element is zero, the optical transition is determined by the linear term in the expansion of $\exp[i(\underline{q} \cdot \underline{r})]$ giving rise to so-called electric quadrupole and magnetic dipole transitions. Compared to the the electric dipole transitions, they are reduced in strength by a factor of $(a/\lambda)^2$.

1.1. Joint density of states

The summation in Eq. 10.16 can be converted into an integration using the methods described in Chapter 5. The resulting integral is very similar to the definition of the density of states. Assuming electric dipole transitions, the energy terms of the valence and conduction band can be combined to

$$E(\underline{k}) = E_G + \frac{\hbar^2}{2m_r^*}\,\underline{k}^2\,, \tag{10.17}$$

since $\underline{k}_h$=$\underline{k}_e$. The dispersion now contains the reduced effective mass m_r^*, which is defined as $m_r^{*^{-1}} = m_e^{*^{-1}} + m_h^{*^{-1}}$. The summation in Eq. 10.16 denoted as the joint density of states depends on the dispersion of both, the conduction and the valence band. The integration leads to

$$g_{3D}\sqrt{\hbar\omega - E_G}\,\Theta(\hbar\omega - E_G)\,, \tag{10.18}$$

where g_{3D} from Eq. 5.16 now contains the reduced effective mass, i.e.,

$$g_{3D} = \frac{(2\,m_r^*)^{3/2}}{2\,\pi^2\,\hbar^3}\,. \tag{10.19}$$

The joint density of states (JDOS) exhibits the same type of critical points as the density of states (DOS) discussed in Chapter 5. The only difference between conventional DOS and JDOS is the appearance of an effective mass of electrons or holes in the DOS, while the JDOS contains the reduced effective mass of the involved bands.

1.2. Direct energy gap

For semiconductors with a direct fundamental energy gap, the optical transitions take place near the center of the Brillouin zone. Assuming a parabolic dispersion, the coefficient g_{3D} in Eq. 10.19 is determined by the reduced effective mass of the valence and conduction band. In this case, the absorption coefficient is given by

$$\begin{aligned}\alpha(\hbar\omega) &= \frac{e^2\sqrt{m}}{\sqrt{2}\,\pi\,\varepsilon_0\,c\,n_r\,\hbar^2}\left(\frac{m_r^*}{m}\right)^{3/2} f_{VC} \\ &\quad\times\sqrt{\hbar\omega - E_G}\,\Theta(\hbar\omega - E_G)\,. \end{aligned} \tag{10.20}$$

This equation can be rearranged as

$$\alpha(\hbar\omega) = \alpha_0^{3D}\sqrt{E_G}\,\sqrt{E^* - 1}\,\Theta(E^* - 1)\,, \tag{10.21}$$

where $\alpha_0^{3D} = A \left(\frac{m_r^*}{m}\right)^{3/2} \frac{f_{VC}}{n_r}$, $E^* = \hbar\omega/E_G$, and $A = e^2\sqrt{m}/(\sqrt{2}\,\pi\,\varepsilon_0\,c\,\hbar^2)$. If the energy gap is measured in eV, the constant A has a value of 7.48×10^5 cm^{-1}. For GaAs 100 meV above the energy gap, the absorption coefficient has a value of 4.74×10^3 cm^{-1} assuming $f_{VC} = 5.3$ obtained from Eq. 10.15 using $\Delta = 0.34$ eV, $n_r = 3.5$, and using the energy gap as well as effective masses listed in Tab. 4.2. If the heavy- and light-hole bands are degenerate at $\underline{k} = 0$, their contribution has to be added together. The total absorption coefficient is therefore the sum of the contribution from the heavy- and light-hole valence bands.

In three-dimensional, direct-gap semiconductors, the absorption at the energy gap E_G is zero and then increases as the square root of $\hbar\omega - E_G$. In two-dimensional systems, the DOS leaps from zero to a finite value at the energy gap. The two-dimensional absorption coefficient therefore also changes discontinuously from 0 to a finite value at the energy gap and remains constant above the energy gap. In one dimension, the DOS and also the JDOS diverge at the energy gap so that the absorption coefficient also diverges in the vicinity of the energy gap. We will return to a more detailed discussion of the absorption properties of lower dimensional systems in the third section of this chapter.

1.3. Indirect energy gap

For semiconductors with an indirect energy gap, the absorption of light in the vicinity of the energy gap involves a large wavevector phonon, which supplies the missing momentum. In this case, the absorption process can be described by a direct optical transition followed by the emission or absorption of a large wavevector phonon, or by the emission or absorption of a phonon followed by a direct optical transition. Energy conservation requires that $\hbar\omega = E_{CV} \pm E_{ph}$, where E_{ph} denotes the phonon energy. Momentum conservation determines the wavevector $\underline{Q}$ of the phonon $\underline{Q} = \underline{k}_C - \underline{k}_V$. In this case, the transition rate has to be obtained from second order perturbation theory by taking into account the combined interaction of electrons with photons and electrons with phonons. The transition rate is given by

$$R(\hbar\omega) = \frac{2\pi}{\hbar}\sum_{\underline{k}_C,\underline{k}_V}\left|\sum_i \frac{\langle C|H_{er}|i\rangle\langle i|H_{ep}|V\rangle}{E_{iv} - \hbar\omega}\right|^2 \times \delta(E_C(\underline{k}_C) - E_V(\underline{k}_V) - \hbar\omega \pm E_{ph}) \,, \quad (10.22)$$

where H_{ep} denotes the electron-phonon interaction Hamiltonian. Assuming that the matrix element does not depend on the wavevector, the summation in Eq. 10.22 can be performed in the continuum limit by replacing it by integrals. Assuming an isotropic band structure, the transition probability becomes proportional to the product of the density of states of the initial and final states and the δ-function ensuring energy conservation. The absorption coefficient can thus be expressed as

$$\alpha(\hbar\omega) \propto \int\int dE_V dE_C g_V(E_V) g_C(E_C) \delta(E_C - E_V - \hbar\omega \pm E_{ph}) \,. \quad (10.23)$$

Using the density of states for the valence and conduction band from Chapter 5, the first integration leads to

$$\alpha(\hbar\omega) \propto \int_{-\infty}^{\infty} dE_C\,(E_C - E_G)^{1/2}\,(\hbar\omega \mp E_{ph} - E_C)^{1/2} \times \Theta(E_C - E_G)\,\Theta(\hbar\omega \mp E_{ph} - E_C) \,. \quad (10.24)$$

Changing the variable in the integral to

$$x = \frac{E_C - E_G}{\hbar\omega \mp E_{ph} - E_G} \quad (10.25)$$

and using the unit step function Θ, the integral can be solved

$$\alpha(\hbar\omega) \propto (\hbar\omega \mp E_{ph} - E_G)^2\,\Theta(\hbar\omega \mp E_{ph} - E_G) \times \int_0^1 dx\; x^{1/2}\,(1 - x)^{1/2} \,. \quad (10.26)$$

The integral has a value of $\pi/8$. Therefore, the energy dependence of the absorption exhibits a quadratic power law for indirect energy-gap semiconductors

$$\alpha(\hbar\omega) \propto (\hbar\omega \mp E_{ph} - E_G)^2\,\Theta(\hbar\omega \mp E_{ph} - E_G) \quad (10.27)$$

instead of the square-root dependence for direct-gap semiconductors. The indirect gap gives rise to two absorption edges, one through absorption of phonons at $\hbar\omega = E_G - E_{ph}$ and another one through emission of phonons at $\hbar\omega = E_G + E_{ph}$. At low temperatures, the density of thermally excited phonons becomes very small, and only the absorption edge due to emission of phonons remains. The two absorption edges can be identified through their different temperature dependence. The indirect gap energy as well as the phonon energy can be determined from the temperature dependence.

1.4. Experimental determination of the optical energy gap

In order to determine the energy gap of a semiconductor, one normally measures the reflection R and transmission T of the respective material as a function of energy. The absorption coefficient can be obtained from the absorbed intensity, which, according to Eq. 10.10, is calculated from

$$\frac{dI}{dx} = -\alpha I . \tag{10.28}$$

Neglecting interference effects, the absorbed intensity normalized to the incident intensity is given by

$$(1 - R)[1 - \exp(-\alpha d)] , \tag{10.29}$$

where d denotes the thickness of the investigated slab. Due to the conservation of energy flux, the sum of the reflected, transmitted, and absorbed intensity should be equal to the incident intensity, i.e.,

$$R + T + (1 - R)[1 - \exp(-\alpha d)] = 1 . \tag{10.30}$$

Solving this equation for α, one obtains

$$\alpha = \frac{1}{d} \ln\left(\frac{1-R}{T}\right) . \tag{10.31}$$

Plotting either α^2 for a direct gap semiconductor or $\sqrt{\alpha}$ for an indirect semiconductor as a function of energy and neglecting any energy dependence of the oscillator strength f_{VC} and the refractive index n_r, a straight line should be obtained. The interception with the energy axis leads to a value for the energy gap. For a direct-gap semiconductor, the slope can be used to determine either the reduced effective mass or the oscillator strength.

2. Absorption Including Excitonic Effects

For photon energies below E_G, the exciton states form a discrete set so that the joint density of states reduces to a sum over δ-functions. The matrix element in Eq. 10.6 has now to be calculated using the ground state as the initial state and the exciton state as the final state taking into account the interaction of the exciton with radiation

$$\langle ex|H_{exr}|0\rangle \,. \tag{10.32}$$

The total wave function of the exciton is the product of the hydrogen-like wave function $\Psi_{nlm}(\underline{r})$ of the relative motion of the exciton and the wave function of electron and holes in the lattice, i.e., the plane wave of the electron $e^{-i\underline{k}\cdot\underline{r}_e}$ times the Bloch wave function of the electron $\psi_{-\underline{k}}(\underline{r}_e)$ and the plane wave of the holes $e^{i\underline{k}\cdot\underline{r}_h}$ times the Bloch wave function of the hole $\psi_{\underline{k}}(\underline{r}_h)$. Neglecting again the photon wavevector, the total wave function of the final state can be written as

$$|ex\rangle \;=\; \sum_{\underline{r},\underline{k}} e^{i\underline{k}\cdot\underline{r}}\,\Psi_{nlm}(\underline{r})\,|\psi_{-\underline{k}}(\underline{r}_e)\,\psi_{\underline{k}}(\underline{r}_h)\rangle \,. \tag{10.33}$$

Using this wave function in Eq. 10.32 and assuming that the matrix element does not depend on $\underline{k}$, one obtains

$$\langle ex|H_{exr}|0\rangle \;=\; \langle C|H_{er}|V\rangle \sum_{\underline{r},\underline{k}} e^{i\underline{k}\cdot\underline{r}}\,\Psi_{nlm}(\underline{r}) \,. \tag{10.34}$$

The summation over $\underline{k}$ results in $\delta(r)$ so that the matrix element is of the form

$$\langle ex|H_{exr}|0\rangle \;=\; \langle C|H_{er}|V\rangle\,\Psi_{nlm}(\underline{0}) \,. \tag{10.35}$$

The square of the absolute value of this matrix element enters into the transition probability. Since for the hydrogen atom, $|\Psi_{nlm}(\underline{0})|$ is only non-zero for s-like states, i.e., $l=0$, only excitons with s symmetry can be excited. The absolute value for arbitrary n can be determined from the Laguerre polynomials of the s-states of the hydrogen wave functions. The volume of the exciton wave function in the nth eigenstate is proportional to the exciton Bohr radius to the third power. In accordance with Eq. 9.8, the absolute value of the wave function at the origin is given by

$$|\Psi_{n00}(\underline{0})|^2 = \frac{1}{\pi\, a^{*3}\, n^3}\,. \tag{10.36}$$

The absorption coefficient for excitonic states below the band-gap energy E_G can therefore be written as

$$\alpha(\hbar\omega) = \frac{m^2 e^8}{128\, c\, \pi^3\, \varepsilon_0^4\, \hbar^5}\left(\frac{m_r^*}{m}\right)^3 \frac{f_{VC}}{n_r^7} \times \sum_n \frac{2}{n^3}\,\delta(\hbar\omega - E_G + E_n)\,. \tag{10.37}$$

A factor of 2 has been included to account for the spin degeneracy. This equation can be rewritten in terms of the exciton binding energy R^* as

$$\alpha(\hbar\omega) = \alpha_0^{3D}\, 2\pi\, \sqrt{R^*}\, \sum_n \frac{2\, R^*}{n^3}\,\delta(\hbar\omega - E_G + E_n)\,. \tag{10.38}$$

Due to the finite line width of the excitonic transitions, the exciton lines overlap for large values of n forming a quasi-continuum just below the energy gap. The density of states in this limit can be approximated by

$$g(E) = \frac{dn}{dE} = \frac{n^3}{2\, R^*}\,. \tag{10.39}$$

As a consequence, the absorption coefficient at the band edge has actually a non-vanishing value of

$$\alpha(\hbar\omega) = \alpha_0^{3D}\, 2\,\pi\, \sqrt{R^*}\,, \tag{10.40}$$

where α_0^{3D} is defined as in Eq. 10.21. This is in strong contrast to the absorption coefficient without excitonic effects.

The continuum states of the exciton above the energy gap also modify the absorption coefficient. In this energy range, the absolute value of the wave function can be expressed as

$$|\Psi_{\eta 00}(\underline{0})|^2 = \frac{\pi\,\eta\,\exp(\pi\,\eta)}{\sinh(\pi\,\eta)}\,, \tag{10.41}$$

where $\eta = \sqrt{R^*/(\hbar\omega - E_G)}$. Multiplying the absorption coefficient without excitonic effects by this quantity leads, for $\hbar\omega > E_G$, to

$$\alpha(\hbar\omega) = \alpha_0^{3D} \frac{2\pi\sqrt{R^*}\,\Theta(\hbar\omega - E_G)}{1 - \exp(-2\pi\sqrt{R^*}/\sqrt{\hbar\omega - E_G})}\,. \tag{10.42}$$

The factor for the spin degeneracy is already included in the absorption coefficient without excitonic effects. For $\hbar\omega = E_G$, this equation reduces to Eq. 10.40. The expression in Eq. 10.41 is called the Sommerfeld or Coulomb enhancement factor, since it describes the change of the free-carrier absorption coefficient due to excitonic effects. The total expression for the absorption coefficient including excitonic effects is given by

$$\alpha(\hbar\omega) = \alpha_0^{3D}\, 2\pi\sqrt{R^*}\left[\sum_n \frac{2}{n^3}\,\delta[(\hbar\omega - E_G)/R^* + 1/n^2] + \frac{\Theta[(\hbar\omega - E_G)/R^*]}{1 - \exp(-2\pi\sqrt{R^*/(\hbar\omega - E_G)})}\right]. \tag{10.43}$$

The total excitonic absorption coefficient is shown in Fig. 10.1 using the line shape function $\delta(x) = \exp(-x^2/\Gamma^2)/(\sqrt{\pi}\,\Gamma)$ for excitons with a binding energy of $R^* = 10$ meV and a line width parameter Γ of $R^*/25 = 0.40$ meV. Note that α already reaches the continuum value below the energy gap in an energy range comparable to Γ. In order to calculate the excitonic absorption spectrum given by Eq. 10.43 for a finite Γ correctly, the continuum states have also been broadened by a Gaussian line shape function. Since the expression for the continuum absorption is equal to unity for $\hbar\omega - E_G \ll 4\pi^2 R^*$, the integral over the line shape function times the continuum absorption expression reduces to the integral over the line shape function. For the Gaussian line shape, this integral becomes $(1 + \mathrm{erf}[(\hbar\omega - E_G)/\Gamma])/2$, which has the form of a broadened step function. The excitonic absorption spectrum for a Gaussian broadening of the excitonic states can therefore be written as

$$\frac{\alpha(\hbar\omega)}{\alpha_0^{3D}} = 2\pi\sqrt{R^*}\left[\sum_n \frac{2R^*}{\sqrt{\pi}\,\Gamma\, n^3}\exp[-(\hbar\omega - E_G + R^*/n^2)^2/\Gamma^2] + \frac{1}{2}\,\frac{1 + \mathrm{erf}[(\hbar\omega - E_G)/\Gamma]}{1 - \exp(-2\pi\sqrt{R^*/|\hbar\omega - E_G|})}\right]. \tag{10.44}$$

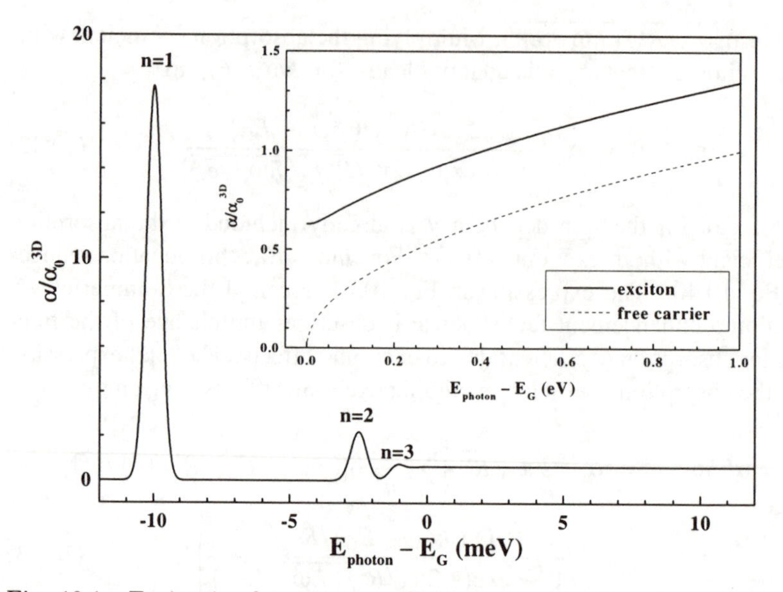

Fig. 10.1. Excitonic absorption coefficient using the line shape function described in the text for the discrete transitions below the energy gap. The inset shows a comparison of the free-carrier absorption (dashed line) with the excitonic absorption (solid line) for energies above E_G.

For other line shape functions such as a Lorentzian or $1/\cosh(x)$, similar arguments apply, and analogous expressions can be obtained.

The inset of Fig. 10.1 shows a comparison of the free-carrier absorption coefficient (dashed line) and the excitonic absorption coefficient for energies above the band gap. The Sommerfeld enhancement is clearly visible even for energies of 1 eV above the energy gap. It approaches the free-carrier absorption coefficient in the limit of very large energies. The ratio of the continuum absorption just above the energy gap and the ground exciton state absorption is given by $\sqrt{\pi}\Gamma/(2R^*)$ so that the continuum absorption is usually more than an order of magnitude smaller than the absorption of the discrete states.

3. Absorption Coefficient in Lower Dimensions

3.1. Free-carrier absorption coefficient

The absorption coefficient without excitonic effects in 2 and 1 dimensions can be directly derived using Eq. 10.16. The joint density of states in this equation has to be replaced by the JDOS of the appropriate dimension. In two dimensions, we obtain with the help of Eq. 5.11

$$\alpha_{2D}(\hbar\omega) \quad = \quad \alpha_0^{2D}\Theta(\hbar\omega - E_G)\,, \tag{10.45}$$

where $\alpha_0^{2D} = A_{2D}\,\frac{m_r^*}{m}\,\frac{f_{VC}}{n_r}$ defines an absorption probability with $A_{2D} = e^2/(2\,\varepsilon_0\,c\,\hbar)$ being a dimensionless constant with a value of 4.59×10^{-2}. The absorption coefficient in two dimensions leaps from zero to a finite value at the energy gap and remains constant above the energy gap.

In one dimension, the absorption coefficient excluding excitonic effects can be determined from Eq. 10.16 and the 1D density of states in Eq. 5.12. The result is

$$\alpha_{1D}(\hbar\omega) \quad = \quad \alpha_0^{1D}\,\frac{\Theta(E^* - 1)}{\sqrt{E_G(E^* - 1)}}\,, \tag{10.46}$$

where $\alpha_0^{1D} = A_{1D}\left(\frac{m_r^*}{m}\right)^{1/2}\frac{f_{VC}}{n_r}$ defines an absorption length with $A_{1D} = e^2/(\sqrt{2m}\,\varepsilon_0\,c)$ being a constant with a value of 1.79×10^{-9} cm, when the energy gap is measured in eV, and $E^* = \hbar\omega/E_G$. In one dimension, the absorption coefficient diverges at the energy gap, when excitonic effects are not taken into account. The energy dependence of the free-carrier absorption coefficient in three, two, and one dimension are displayed in Fig. 10.2 using different confinement energies in two and one dimensions.

3.2. Excitonic absorption coefficient

The inclusion of excitonic effects can be easily performed for two dimensions. The exciton energies and the respective Bohr radius were al-

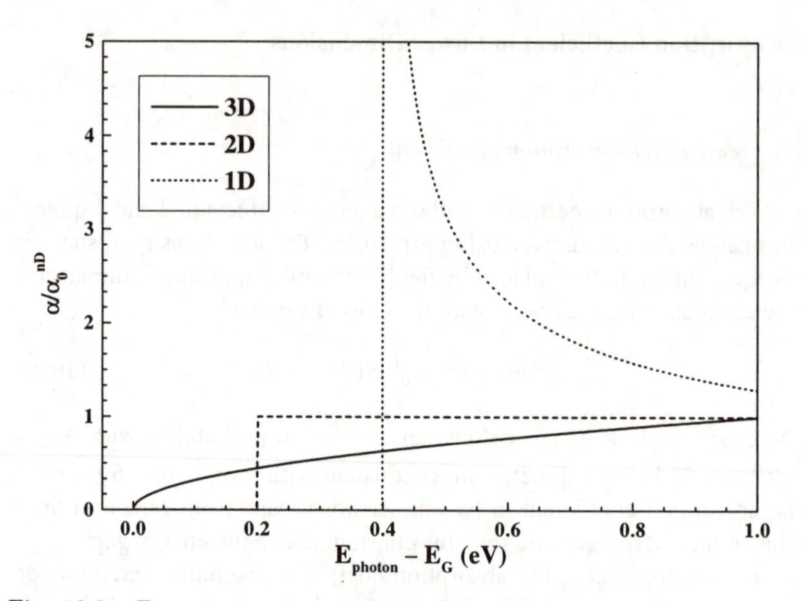

Fig. 10.2. Free-carrier absorption coefficient for three-, two-, and one-dimensional systems. The energy gap in two and one dimensions is increased due to confinement effects. The strength of the absorption coefficient is not comparable between the different dimensions.

ready given in Eqs. 9.9 and 9.10, respectively. The procedure is very similar to the calculation for three dimensions. In two dimensions, the value for the wave function overlap at the origin is given by

$$|\Psi_{n0}^{2D}(\underline{0})|^2 \;=\; \frac{1}{\pi\, a^{*2}\,(n-1/2)^3}\,. \qquad (10.47)$$

Using again Eqs. 10.6 and 10.11, one obtains for energies below the energy gap

$$\alpha_{2D}(\,\hbar\omega) \;=\; 2\,\alpha_0^{2D} \sum_n \frac{2\,R^*}{(n-1/2)^3}\,\delta(\,\hbar\omega\,-E_G\,+\,E_n)\,. \qquad (10.48)$$

A factor of 2 for spin degeneracy has been included. The reduction of the amplitude of higher exciton states is considerably larger in two dimensions than in three. For example, the ratio of the amplitude of the first and second exciton state in three dimensions is 1/8, while in two dimensions it is 1/27. However, the relative amplitude of the ground state is higher, since the ground state amplitude is 1 in 3D and 8 in 2D. The quasi-continuum just below the energy gap can be treated in the same way as in 3D. The density of states is given by

$$g_{2D}(E) = \frac{dn}{dE} = \frac{(n-1/2)^3}{2\,R^*} . \tag{10.49}$$

Therefore, the absorption coefficient at the band edge has actually a value of

$$\alpha_{2D}(\hbar\omega) = 2\,\alpha_0^{2D} , \tag{10.50}$$

i.e., it is a factor of 2 larger than the absorption coefficient without excitonic effects. While in three dimensions the absorption coefficient at $\hbar\omega = E_G$ increases from zero to a finite value, when excitonic effects are included, the absorption coefficient in two dimensions increases at $\hbar\omega = E_G$ only by a factor of two.

The continuum states in 2D can be treated similarly as in 3D. The value for the wave function overlap at the origin is given by

$$|\Psi_{\eta 0}^{2D}(\underline{0})|^2 = \frac{\exp(\pi\,\eta)}{\cosh(\pi\,\eta)} , \tag{10.51}$$

where $\eta^2 = R^*/(\hbar\omega - E_G)$. The total two-dimensional absorption coefficient including excitonic effects has therefore the form

$$\alpha_{2D}(\hbar\omega) = 2\,\alpha_0^{2D} \left[\sum_n \frac{2}{(n-1/2)^3}\, \delta\!\left(\frac{\hbar\omega - E_G}{R^*} + \frac{1}{(n-1/2)^2} \right) + \frac{\Theta[(\hbar\omega - E_G)/R^*]}{1 + \exp(-2\pi\sqrt{R^*/(\hbar\omega - E_G)})} \right] . \tag{10.52}$$

The factor for spin degeneracy is already included in the two-dimensional absorption coefficient without excitonic effects. The excitonic absorption coefficient for a two-dimensional system is shown in Fig. 10.3. The numerical calculation of the excitonic spectrum for 2D was performed in the same way as for 3D using the same parameters $R^* = 10$ meV and

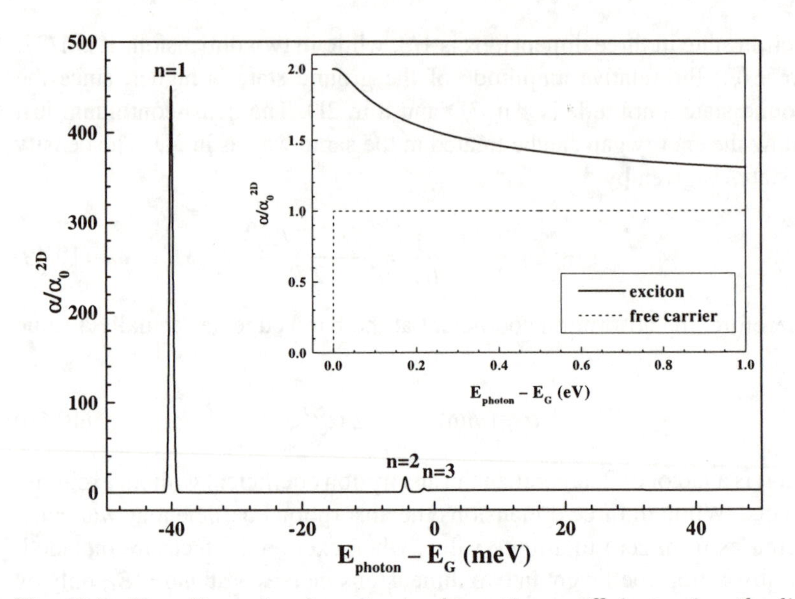

Fig. 10.3. Two-dimensional excitonic absorption coefficient using the line shape function described in the text for the discrete transitions below the energy gap. The inset shows a comparison of the free-carrier absorption (dashed line) with the excitonic absorption (solid line) for energies above E_G.

$\Gamma = R^*/25 = 0.4$ meV. Note that the excitonic binding energy is $4R^*$. The inset shows a comparison of the free-carrier (dashed line) with the exciton absorption (solid line) for energies above the energy gap. The Sommerfeld enhancement factor is again clearly visible in the vicinity of the energy gap with a value of 2 at the energy gap. It reduces to one in the limit of very large energies. The ratio of the continuum absorption just above the energy gap and the ground exciton state absorption is $\sqrt{\pi}\Gamma/(16R^*)$ so that the continuum absorption in high-quality crystals is about two orders of magnitude smaller than the dominant exciton absorption.

The calculation of the change in the one-dimensional absorption spectrum due to excitonic effects is rather involved. We will therefore only note that the Sommerfeld enhancement factor in 1D actually results in a reduction of the absorption coefficient of free carriers in strong contrast to two and three dimensions. The singular density of states of free carriers in 1D does not appear in the absorption spectrum. Therefore, the spectral features at the band gap of 1D systems are actually less pronounced than in 2D- and 3D-systems, and consequently it is rather difficult to determine the energy gap of such systems exactly.

4. Absorption at Finite Temperatures

So far, we have only considered the absorption coefficient at zero temperature, i.e., no electrons or holes were thermally excited across the energy gap. In order to take finite temperature effects into account, the absorption coefficient at zero temperature, which will be denoted α_0, has to be multiplied with the product of the density of unoccupied electron states in the conduction band and that of occupied electron states in the valence band. The number of unoccupied electron states in the conduction band is given by

$$1 - f_e(\frac{\hbar\omega}{k_B T}) , \tag{10.53}$$

while the number of occupied electron states in the valence band is equal to the number of unoccupied hole states, i.e.,

$$1 - f_h(\frac{\hbar\omega}{k_B T}) . \tag{10.54}$$

Furthermore, we have to take into account the contribution from stimulated emission. However, since the transition probabilities for absorption and stimulated emission are the same, the gain coefficient for stimulated emission is equal to $-\alpha_0(\hbar\omega)$, since the transition probabilities are equal. Stimulated emission can only take place, when electron and hole states are occupied. We therefore obtain for the absorption coefficient at finite temperature

$$\begin{aligned} \alpha(\hbar\omega, T) &= \alpha_0(\hbar\omega)\,[1 - f_h(\hbar\omega, T)]\,[1 - f_e(\hbar\omega, T)] \\ &\quad - \alpha_0(\hbar\omega)\, f_e(\hbar\omega, T)\, f_h(\hbar\omega, T) . \end{aligned} \tag{10.55}$$

The resulting absorption coefficient at finite temperature has the simple form

$$\alpha(\hbar\omega, T) = \alpha_0(\hbar\omega)\,[1 - f_h(\hbar\omega, T) - f_e(\hbar\omega, T)] . \tag{10.56}$$

This equation can be used for any dimension. However, since the effective masses for electron and holes are usually different, the energy term in the Fermi-Dirac distribution function has to be evaluated with care. Finally, we have to mention that the energy gap of semiconductors exhibits a shift with increasing temperature, which is typically a red-shift. This energy shift occurs due to the increased anharmonicity of lattice vibrations with increasing temperature.

5. Emission of Light

5.1. Excess energy

In this section, we will briefly discuss the inverse process of absorption, i.e., light emission. After excitation of a certain carrier density by a light source, the electrons and holes will thermalize to the respective band edge. This process usually occurs on a very short time scale (within a few picoseconds). For a photon energy of $\hbar\omega$, the excess energy after photoexcitation for electrons is given by

$$\begin{aligned} E^e_{exc}(\hbar\omega) &= \frac{\hbar^2 k^2}{2\, m_e^*} = \frac{m_h^*}{m_e^* + m_h^*}\,(\hbar\omega - E_G) \\ &= \frac{m_r^*}{m_e^*}\,(\hbar\omega - E_G)\,. \end{aligned} \tag{10.57}$$

The corresponding expression for holes is given by

$$\begin{aligned} E^h_{exc}(\hbar\omega) &= \frac{\hbar^2 k^2}{2\, m_h^*} = \frac{m_e^*}{m_e^* + m_h^*}\,(\hbar\omega - E_G) \\ &= \frac{m_r^*}{m_h^*}\,(\hbar\omega - E_G)\,. \end{aligned} \tag{10.58}$$

Since the effective mass of the heavy holes is usually much larger than the electron effective mass, the excess energy of the electrons is much larger. In GaAs, the relative excess energy is 85% for electrons and 15% for heavy holes. The thermalization process to the band edge is dominated by the emission of optic and acoustic phonons.

5.2. Spontaneous emission

The thermalized electron-hole pairs can recombine radiatively giving rise to light emission. There are two important factors to consider. The first one, which was discussed in the previous section, is the thermal occupation. Emission can only take place between a filled electron state in the conduction band and an empty state in the valence band, i.e., an occupied hole state. The second one is the probability for spontaneous emission. The rate of recombination per unit volume, unit time, and unit energy is given by the product of the Einstein coefficient A_{CV} for spontaneous emission, the density of occupied conduction band states n_C, and the density of unoccupied valence band states n'_V

$$R_{CV}(\hbar\omega) = A_{CV}(\hbar\omega)\, n_C\, n'_V\,. \qquad (10.59)$$

The Einstein coefficient $A_{CV}(\hbar\omega)$ is related to the Einstein coefficient for stimulated emission $B_{CV}(\hbar\omega)$ via

$$A_{CV}(\hbar\omega) = \frac{n_r^3\,\omega^3}{\pi^2\,c^3}\, B_{CV}(\hbar\omega)\,. \qquad (10.60)$$

Since $B_{CV}(\hbar\omega)$ is equal to the Einstein coefficient for absorption $B_{VC}(\hbar\omega)$, the recombination rate can be directly related to the absorption coefficient. The transition rate for absorption per unit volume, unit time, and unit energy given in Eq. 10.8 is connected with B_{VC} via

$$R(\hbar\omega) = B_{VC}(\hbar\omega)\,\frac{n_r}{c}\, I\,(n_V\, n'_C - n_C\, n'_V)\,. \qquad (10.61)$$

The first term corresponds to absorption, the second to stimulated emission. Since according to Eq. 10.11 $R(\hbar\omega) = I\,\alpha(\hbar\omega)/\hbar\omega$, the relation with the absorption coefficient becomes

$$\alpha(\hbar\omega, T) = B_{VC}(\hbar\omega)\,\frac{n_r\,\hbar\omega}{c}\,(n_V\, n'_C - n_C\, n'_V)\,. \qquad (10.62)$$

Combining this equation with the one for spontaneous emission results in the well-known Van Roosbroeck-Shockley relation, which connects the recombination rate with the absorption coefficient

$$R_{CV}(\hbar\omega) = \frac{\alpha(\hbar\omega)}{\hbar}\left(\frac{n_r\,\omega}{\pi\,c}\right)^2 \frac{n_C\, n'_V}{n_V\, n'_C - n_C\, n'_V}\,. \qquad (10.63)$$

For band-to-band transitions in an intrinsic semiconductor in equilibrium, it can be shown using the Fermi-Dirac distribution functions that the last term in Eq. 10.63 reduces to the Bose-Einstein distribution function for photons, i.e.,

$$R_{CV}(\hbar\omega) \;=\; \frac{\alpha(\hbar\omega)}{\hbar}\left(\frac{n_r\,\omega}{\pi\,c}\right)^2 \frac{1}{\exp(\hbar\omega/k_BT)-1}\,. \qquad (10.64)$$

Since in most cases $\hbar\omega \gg k_BT$, the Bose-Einstein distribution function can be approximated by the corresponding Maxwell-Boltzmann distribution so that

$$R_{CV}(\hbar\omega) \;=\; \frac{\alpha(\hbar\omega)}{\hbar}\left(\frac{n_r\,\omega}{\pi\,c}\right)^2 \exp(-\hbar\omega/k_BT)\,. \qquad (10.65)$$

Integrating this expression over energy results in the transition probability per unit volume and time

$$R_{CV} \;=\; \frac{1}{\hbar^3\,\pi^2\,c^2}\int_0^\infty d(\hbar\omega)\,\frac{n_r^2\,(\hbar\,\omega)^2\,\alpha(\hbar\omega)}{\exp(\hbar\omega/k_BT)-1}\,. \qquad (10.66)$$

Generally, the photon energy in the exponential function has to be replaced by $\hbar\omega - \Delta E_F$, where ΔE_F denotes the difference of the quasi-Fermi levels of electrons in the conduction and holes in the valence band. The expression in Eq. 10.63 can also be used for non-equilibrium conditions as well as transitions between localized and extended or only localized states.

The emitted photoluminescence intensity I_{PL} per unit energy can be directly obtained from the rate of recombination, the electron and hole densities, the emitted photon energy, and the absorption length at the excitation energy $\hbar\omega_{ex}$

$$I_{PL}(\hbar\omega) \;=\; \frac{\hbar\omega}{\alpha(\hbar\omega_{ex})}R_{CV}(\hbar\omega)\,. \qquad (10.67)$$

Integrating this expression over energy results in the emitted intensity in units of W m^{-2}. For an intrinsic semiconductor in equilibrium, the spectral shape of the PL signal can be described by the following expression

$$I_{PL}(\hbar\omega) \;=\; \frac{\alpha(\hbar\omega)}{\alpha(\hbar\omega_{ex})}\,\frac{n_r^2\,\omega^3}{\pi^2\,c^2}\,n\,p\,\exp(-\hbar\omega/k_BT)\,. \qquad (10.68)$$

For a three-dimensional semiconductor, the absorption coefficient for free carriers can be taken from Eq. 10.21 so that Eq. 10.68 becomes

$$I_{PL}(\hbar\omega) = \frac{\alpha_0^{3D}\, n_r^2}{\alpha(\hbar\omega_{ex})\, \pi^2\, \hbar^3\, c^2}\, n\, p\, \sqrt{\hbar\omega - E_G}\, (\hbar\omega)^3 \times \Theta(\hbar\omega - E_G)\exp(-\hbar\omega/k_B T)\,. \qquad (10.69)$$

The product of the step function and the exponential function confines the emitted PL signal to a rather small energy range in contrast to the absorption signal. In Fig. 10.4, the calculated emission spectrum is shown for two temperatures in comparison with the absorption spectrum. While the absorption spectrum is hardly affected by a temperature change between 10 and 300 K for an energy gap of 1.5 eV, the emission spectrum exhibits a strong change in the line width. The strength of the PL signal has been normalized to its maximum value. The high-energy tail of the emission spectrum can be used to determine the carrier temperature, since its slope

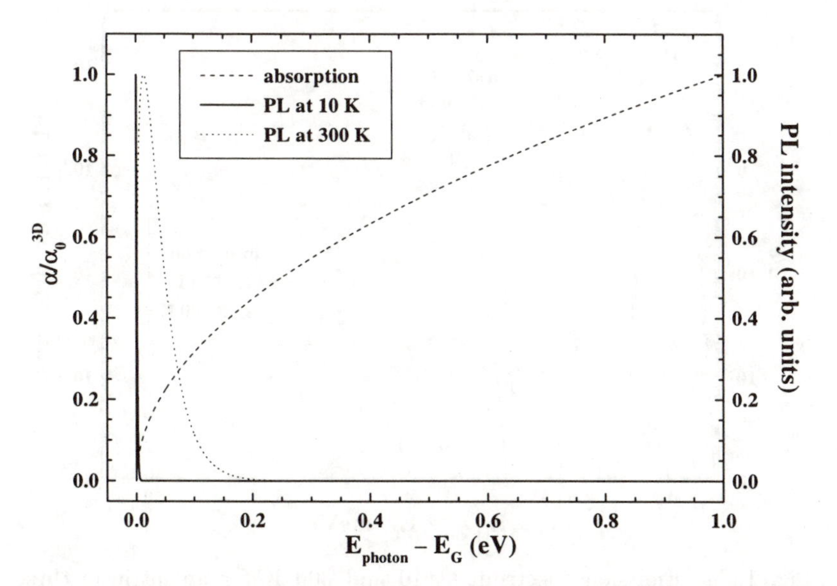

Fig. 10.4. Emission spectrum at 10 and 300 K for an intrinsic three-dimensional semiconductor with an energy gap of 1.5 eV for band-to-band transitions.

on a logarithmic scale is approximately proportional to the inverse temperature. However, the energy-dependent prefactors should also be considered to unambiguously identify the carrier temperature. For an intrinsic semiconductor, the spectrally integrated PL intensity and also the intensity at the maximum of the line increases with increasing temperature, since more and more carriers are thermally excited across the energy gap. However, for a photoexcited semiconductor, the spectrally integrated PL signal is constant, since the carrier density is determined by the intensity of the exciting light source. Since the lines become broader with increasing temperature, the PL intensity at the maximum of the line decreases with increasing temperature.

For excitonic transitions, the Van Roosbroeck-Shockley relation can also be used to deduce the emission spectrum. Since only the absorption coefficient is different, we can use Eq. 10.68 to calculate the excitonic PL signal. In Fig. 10.5, the calculated emission spectrum for excitons is

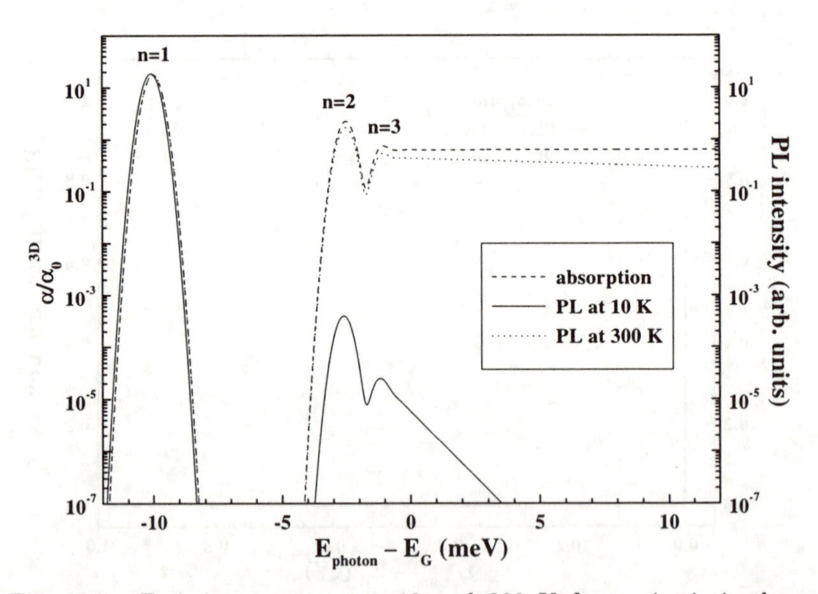

Fig. 10.5. Emission spectrum at 10 and 300 K for an intrinsic three-dimensional semiconductor with an energy gap of 1.5 eV for excitonic transitions.

shown for two temperatures in comparison with the corresponding absorption spectrum shown in Fig. 10.1. The line width has been assumed to remain constant between 10 and 300 K. The PL intensity has been normalized to unity at the maximum of the exciton ground state. At 300 K, the difference between absorption and emission spectrum can only be seen for the higher exciton and the continuum states. However, at 10 K, the intensity of the higher exciton and the continuum states is strongly reduced with respect to the ground state. The approximate exponential dependence is clearly visible for the continuum states. The spectral position of the exciton ground state is also affected at low temperatures. The PL line is shifted to lower energies with respect to the absorption spectrum by about 0.1 meV. For a Gaussian line, this shift is given by

$$\Delta E_{max} = -\frac{\Gamma^2}{2\,k_B T}\,. \tag{10.70}$$

Using the applied line width of 0.4 meV, this shift becomes 0.093 meV at 10 K, while at 300 K it is negligible with a value of 0.003 meV. Usually, the line width parameter Γ will also increase with increasing temperature. Assuming that the line width parameter is proportional to $k_B T$, the red-shift of the PL line with respect to the absorption signal will be proportional to $k_B T$.

5.3. Carrier dynamics

The spectrally integrated rate of recombination is an important quantity in the description of the carrier dynamics. For band-to-band transitions, this rate can be written as

$$R_{CV} = B\,n\,p\,. \tag{10.71}$$

The dynamics of carriers is described by the differential equations

$$-\frac{dn(t)}{dt} = -\frac{dp(t)}{dt} = B\,n(t)\,p(t)\,. \tag{10.72}$$

This simple picture results in a nonlinear equation for the time evolution of the electron and hole population. For an intrinsic material, $n = p$ and the integration of Eq. 10.72 is straightforward resulting in

$$n(t) = \frac{n(0)}{1 + B\,n(0)\,t}, \tag{10.73}$$

where $n(0)$ denotes the electron density at $t = 0$. Since in this case the time dependence of both types of carriers enter into the description, one speaks of bimolecular recombination.

For an extrinsic material, where the carrier density is determined by the doping level, the density of photoexcited carriers is usually smaller than the carrier density originating from doping. Assuming a density p_0 originating from doping, which is much larger than the photoexcited carrier density $n_1 = p_1$, Eq. 10.72 becomes

$$\frac{dn_1(t)}{dt} = -B\,n_1(t)\,[p_0 + p_1(t)] \approx -B\,n_1(t)\,p_0 . \tag{10.74}$$

This equation is now linear, and the solution is the well-known exponential function

$$n_1(t) = n_1(0)\exp[-t/(B\,p_0)] . \tag{10.75}$$

Since the time dependence of only one type of carrier enters into the description, this is usually referred to as mono-molecular recombination. In this case, we can directly deduce a carrier lifetime τ given by

$$\tau = \frac{1}{B\,p_0} . \tag{10.76}$$

Defining a generalized time constant as

$$\tau^{-1} = -\frac{1}{n(t)}\frac{dn(t)}{dt}, \tag{10.77}$$

we recover the already given result for mono-molecular recombination. However, for bimolecular recombination the obtained time constant becomes time-dependent, since

$$\tau^{-1} = B\,n = \frac{B\,n(0)}{1 + B\,n(0)\,t} . \tag{10.78}$$

The dynamic behavior of band-to-band transitions in an intrinsic material corresponds to the bimolecular case, while the dynamics of band-to-band transitions in an extrinsic material and of excitonic transitions can be described by mono-molecular recombination. Therefore, exciton recombination usually follows an exponential time dependence with a well-defined time constant. Generalizing Eq. 10.78 to the simultaneous presence of arbitrary equilibrium densities, the radiative lifetime can be obtained from

$$\tau_R^{-1} = B\,[n_0 + p_0 + \delta n(t)]\,, \tag{10.79}$$

where $\delta n(t)$ denotes the deviation of the electron density from its equilibrium value n_0. With the help of Eq. 10.71, we arrive

$$\tau_R = \frac{np}{R_{CV}\,[n_0 + p_0 + \delta n(t)]}\,. \tag{10.80}$$

The maximum radiative lifetime is obtained by setting $\delta n(t) = 0$. Since $R_{CV}\, n_0\, p_0 = R^0_{CV}\, n\, p$,

$$\tau_R^{max} = \frac{n_0\, p_0}{R^0_{CV}\,(n_0 + p_0)}\,. \tag{10.81}$$

For an intrinsic semiconductor, $n_0 = p_0 = n_i$ so that

$$\tau_R^{max} = \frac{n_i}{2\,R^0_{CV}}\,. \tag{10.82}$$

Using the expression for the equilibrium transition rate for spontaneous emission for a non-degenerate semiconductor from Eq. 10.65 and assuming $\hbar\omega \gg k_BT$, the integral in Eq. 10.66 can be approximated by

$$R^0_{CV} = \frac{n_r^2\,\alpha_0^{3D}}{2\,\hbar^3\,\pi^{3/2}\,c^2}\,E_G^2\,(k_BT)^{3/2}\,\exp(-E_G/k_BT)\,. \tag{10.83}$$

The intrinsic carrier density can be taken from Eq. 5.32. The maximum carrier lifetime for intrinsic semiconductors is therefore

$$\tau_R^{max} = \frac{\sqrt{N_C(T)\,N_V(T)}\,2\,\hbar^3\,\pi^{3/2}\,c^2}{n_r^2\,\alpha_0^{3D}\,E_G^2\,(k_BT)^{3/2}}\,\exp[E_G/(2\,k_BT)]\,. \tag{10.84}$$

According to Eq. 5.36, N_C and N_V are proportional to $(k_BT)^{3/2}$. Using the definition of α_0^{3D} from Eq. 10.21, the final expression of the maximum radiative lifetime is of the form

$$\tau_R^{max} = D \frac{(m_e^* + m_h^*)^{3/2}}{(m_e^* m_h^*)^{3/4}} \frac{\exp[E_G/(2 k_BT)]}{n_r f_{VC} E_G^2}, \tag{10.85}$$

where $D = m^{3/2} c^2/A = \pi \varepsilon_0 m c^3 \hbar^2/e^2$ is independent of any material parameters. For GaAs at room temperature, the maximum lifetime according to this equation should have a value of about 2200 s, which is much too large compared to the measured lifetimes in a direct-gap semiconductor such as GaAs.

For an extrinsic semiconductor with a much larger electron than hole density, the radiative lifetime becomes

$$\tau_R = \frac{n_i^2}{R_{CV}^0 n_0}. \tag{10.86}$$

For a p-type semiconductor, n_0 has to be replaced by p_0. Using the expressions for n_i^2 and R_{CV}^0, one obtains

$$\tau_R = D^* \frac{(m_e^* + m_h^*)^{3/2}}{m^{3/2} n_r f_{VC} E_G^2 n_0} (k_BT)^{3/2}, \tag{10.87}$$

where $D^* = m^3c^2/(\pi^{3/2} \hbar^3 A) = \sqrt{2} m^{5/2} \varepsilon_0 c^3/(\sqrt{\pi} e^2 \hbar)$ is again a constant independent of any material parameters. If all donors (acceptors) are ionized, $n_0 = N_d$ ($p_0 = N_a$) so that the lifetime is determined by the density of donors (acceptors). For GaAs at room temperature and a donor density of 10^{15} cm^{-3}, one obtains a lifetime of about 6.5 μs. Note that the radiative lifetime appears to increase with increasing temperature. However, n_0 is also temperature dependent as discussed in Chapter 5. Since the intrinsic carrier density for a typical semiconductor is much smaller than any background doping density, the radiative lifetime in semiconductors is usually determined by the background doping density.

CHAPTER 11
ELECTROABSORPTION

The application of an external electric field breaks the translational symmetry of the lattice in the direction parallel to the field. The change of the eigenenergies of a quantum mechanical system due to an external electric field is known as the (dc) Stark effect. In bulk semiconductors, the influence of the electric field on the absorption coefficient is called the Franz-Keldysh effect. In lower dimensional structures such as quantum wells, the application of an electric field parallel to the layer has a similar effect as in three dimensions. However, when the electric field is applied perpendicular to the layer, the eigenenergies of the quantum well are directly influenced by the Stark effect, which is known as the quantum-confined Stark effect (QCSE), resulting in a quadratic red-shift of the confinement energies.

1. Electroabsorption of Free Carriers

Under a uniform electric field parallel to the z-direction, i.e., $\underline{F} = F\underline{\hat{z}}$, and neglecting any Coulomb effects between the carriers, the Schrödinger equation for the electron-hole pair can be expressed as

$$\left(-\frac{\hbar^2}{2\,m_r^*}\,\Delta_{\underline{r}} - e\,F\,z\right)\Psi(\underline{r}) \;=\; E\,\Psi(\underline{r})\,. \tag{11.1}$$

Separating the motion perpendicular and parallel to the field direction, we obtain for the motion in the z-direction

$$\left(-\frac{\hbar^2}{2\,m_r^*}\,\frac{d^2}{dz^2} - e\,F\,z\right)\Psi(z) \;=\; E_z\,\Psi(z)\,, \tag{11.2}$$

while the in-plane motion of the carriers is the same as in the absence of an electric field, i.e., the wave function is given by

$$\Psi(x,y) \;=\; A\exp\left[i(k_x\,x + k_y\,y)\right]\,, \tag{11.3}$$

where A is a normalization constant. The total energy is given by

$$E = \frac{\hbar^2 \, (k_x^2 + k_y^2)}{2 \, m_r^*} + E_z \, . \tag{11.4}$$

The task is now to determine E_z, which can be obtained by solving the differential equation in Eq. 11.2 using the appropriate boundary condition. This equation can be reduced to the following equation

$$\frac{d^2\Psi(\xi)}{d\xi^2} = - \, \xi \, \Psi(\xi) \, , \tag{11.5}$$

where the dimensionless variable ξ defined as

$$\xi = \frac{E_z}{E(F)} + \frac{z}{z_0} \tag{11.6}$$

has been introduced. The electrooptical energy $E(F)$ is defined as

$$E(F) = \left(\frac{e^2 \, \hbar^2 \, F^2}{2 \, m_r^*} \right)^{1/3} , \tag{11.7}$$

while the coordinate z_0 is determined by

$$z_0 = \left(\frac{\hbar^2}{2 \, m_r^* \, e \, F} \right)^{1/3} . \tag{11.8}$$

The solutions of the differential equation $f''(x) = x f(x)$ in Eq. 11.5 are the Airy functions. There are two independent solutions, $\mathrm{Ai}(x)$ and $\mathrm{Bi}(x)$, but only $\mathrm{Ai}(x)$ fulfills the boundary condition of being finite for $x \to \infty$. It is usually defined through an integral

$$\mathrm{Ai}(x) = \frac{1}{\pi} \int_0^\infty ds \cos \left(\frac{s^3}{3} + s \, x \right) . \tag{11.9}$$

The normalized wave function has the form (note the different sign of the variable)

$$\Psi_n(z) = a_n \, \mathrm{Ai}(-\xi_n) \, . \tag{11.10}$$

The normalization constant a_n is given by $\pi/\sqrt{z_0 L}$ in the limit $L \to \infty$, where L denotes the size of the system. In this limit, the Airy function $\mathrm{Ai}(x)$ has the following asymptotic form

$$\begin{aligned} \lim_{x\to\infty} \mathrm{Ai}(x) &= \frac{1}{2\sqrt{\pi}\, x^{1/4}} \left(1 - \frac{3\,c_1}{2}\, x^{-3/2}\right) \\ &\quad \times \exp\left(-\frac{2}{3}\, x^{3/2}\right), \end{aligned} \tag{11.11}$$

where $c_1 = 15/216$. In the other limit, when $L \to -\infty$, $\mathrm{Ai}(x)$ is an oscillatory function

$$\lim_{x\to\infty} \mathrm{Ai}(-x) = \frac{1}{\sqrt{\pi}\, x^{1/4}} \sin\left(\frac{2}{3}\, x^{3/2} + \frac{\pi}{4}\right). \tag{11.12}$$

The energies E_z are determined by the condition $\Psi(z = L) = 0$. Using the asymptotic form in Eq. 11.12, one obtains

$$E_z = E(F) \left[\left(\frac{3}{2}(n - \frac{1}{4})\,\pi\right)^{2/3} - \frac{L}{z_0}\right]. \tag{11.13}$$

The total energy is given by Eq. 11.4. Introducing the in-plane momentum $k_\parallel^2 = k_x^2 + k_y^2$, the total energy becomes

$$E_{n,k_\parallel} = \frac{\hbar^2\, k_\parallel^2}{2\, m_r^*} - e\,F\,L + \left(\frac{3\, e\, \hbar\, F}{2\sqrt{2\, m_r^*}}(n - \frac{1}{4})\,\pi\right)^{2/3}. \tag{11.14}$$

Using the expression for the absorption coefficient from Chapter 10 and adding the result as in Eq. 10.34 for the wave function at the origin, we obtain the following expression for the absorption coefficient

$$\begin{aligned} \alpha(\hbar\omega) &= \frac{\pi\, e^2\, \hbar}{2\, \varepsilon_0\, m\, c\, n_r}\, f_{VC} \\ &\quad \times \sum_{\underline{k}} |\Psi(0)|^2\, \delta(E_C(\underline{k}) - E_V(\underline{k}) - \hbar\omega). \end{aligned} \tag{11.15}$$

The summation over k_x and k_y can be converted to an integration. The sum over the eigenenergies E_z can also be expressed as an integration using the density of states as derived from the derivative dE/dn.

$$\frac{dn}{dE} = \frac{\left(\frac{3}{2}\left(n-\frac{1}{4}\right)\pi\right)^{1/3}}{\pi\, E(F)} \quad . \tag{11.16}$$

In the limit of $L \to \infty$, Eq. 11.13 can be used to express the term depending on n in Eq. 11.16 according to

$$g(E_z) = \frac{dn}{dE} = \frac{\sqrt{L/z_0}}{\pi\, E(F)} \,. \tag{11.17}$$

The value of the wave function at the origin in Eq. 11.15 is given by

$$|\Psi(z=0)|^2 = \frac{\pi}{\sqrt{z_0\, L}} \left| \mathrm{Ai}\left(-\frac{E_z}{E(F)}\right)\right|^2 , \tag{11.18}$$

where the expression for E_z in Eq. 11.13 has been used. The absorption coefficient divided by the prefactor α_{pre} as defined by

$$\alpha_{pre} = \frac{\pi\, e^2\, \hbar}{2\, \varepsilon_0\, m\, c\, n_r}\, f_{VC} \tag{11.19}$$

can be written as follows

$$\frac{\alpha(\hbar\omega)}{\alpha_{pre}} = \int_0^\infty dE_{k_\|}\, g(E_{k_\|})\, dE_z\, g(E_z)\, |\Psi(z=0)|^2 \times \delta(E_{k_\|} + E_z + E_G - \hbar\omega)\,. \tag{11.20}$$

Since the density of states for E_z does not depend on energy, the integration over E_z removes the δ-function, i.e.,

$$\frac{\alpha(\hbar\omega)}{\alpha_{pre}} = \frac{1}{z_0\, E(F)} \int_0^\infty dE_{k_\|}\, g(E_{k_\|}) \times \left| \mathrm{Ai}\left(\frac{E_{k_\|} + E_G - \hbar\omega}{E(F)}\right)\right|^2 \quad . \tag{11.21}$$

The density of states for the in-plane motion corresponds to the two-dimensional DOS discussed in Chapter 5. It is given by

$$g(E_{k_\|}) = \frac{m_r^*}{\pi\, \hbar^2}\, \Theta(E_{k_\|})\,. \tag{11.22}$$

Using this expression and transforming the variable $E_{k_\parallel}$ to $x = (E_{k_\parallel} + E_G - \hbar\omega)/E(F)$, the absorption coefficient can be expressed as

$$\alpha(\hbar\omega) = \frac{\alpha_{pre}}{z_0} \frac{m_r^*}{\pi \hbar^2} \int_{x_0}^{\infty} dx \, |\mathrm{Ai}(x)|^2 \, , \tag{11.23}$$

where the lower limit of the integral is given by

$$x_0 = \frac{E_G - \hbar\omega}{E(F)} \, . \tag{11.24}$$

The integral in Eq. 11.23 can be rewritten in terms of the Airy function and its derivative using partial integration and the differential equation for the Airy function itself [$\mathrm{Ai}''(x) = x\mathrm{Ai}(x)$]

$$\begin{aligned} \int_{x_0}^{\infty} dx \, |\mathrm{Ai}(x)|^2 &= x \, \mathrm{Ai}^2(x)|_{x_0}^{\infty} - \int_{x_0}^{\infty} dx \, x \, 2 \, \mathrm{Ai}(x) \, \mathrm{Ai}'(x) \\ &= -x_0 \, \mathrm{Ai}^2(x_0) - \int_{x_0}^{\infty} dx \, 2 \, \mathrm{Ai}''(x) \, \mathrm{Ai}'(x) \\ &= -x_0 \, \mathrm{Ai}^2(x_0) + \mathrm{Ai}'^2(x_0) \, . \end{aligned} \tag{11.25}$$

Using the definition of the prefactor in Eq. 10.20, the absorption coefficient becomes

$$\begin{aligned} \alpha(\hbar\omega, F) &= \alpha_0^{3D} \, \pi \, \sqrt{E(F)} \\ &\times \left(-x_0 \, \mathrm{Ai}^2(x_0) + \mathrm{Ai}'^2(x_0) \right) \, . \end{aligned} \tag{11.26}$$

We can evaluate Eq. 11.26 using the asymptotic forms of the Airy functions in Eqs. 11.11 and 11.12. For energies above the energy gap, we have to use the limit $x_0 \to -\infty$, i.e., Eq. 11.12. The derivative of this asymptotic form is given by

$$\begin{aligned} \lim_{x \to \infty} \mathrm{Ai}'(-x) &= \frac{x^{1/4}}{\sqrt{\pi}} \left[\cos\left(\frac{2}{3} x^{3/2} + \frac{\pi}{4} \right) \right. \\ &\left. - \frac{1}{4 \, x^{3/2}} \sin\left(\frac{2}{3} x^{3/2} + \frac{\pi}{4} \right) \right] \, . \end{aligned} \tag{11.27}$$

Since x_0 is negative for energies above the gap, the first term in Eq. 11.26 is positive. We therefore obtain for the absorption coefficient in the limit of low fields or high energies

$$\alpha(\hbar\omega,F) = \alpha_0^{3D}\,\Theta(\hbar\omega - E_G)\,\sqrt{\hbar\omega - E_G}$$
$$\left[1 - \frac{1}{4}\left(\frac{E(F)}{\hbar\omega - E_G}\right)^{3/2}\cos\left(\frac{4}{3}\left(\frac{\hbar\omega - E_G}{E(F)}\right)^{3/2}\right)\right] . \quad (11.28)$$

In this expression, we have only taken into account the leading order in x_0. The absorption coefficient above the energy gap exhibits damped oscillations with a period of

$$T(\hbar\omega) = \frac{3}{4}\frac{E(F)^{3/2}}{\sqrt{\hbar\omega - E_G}} = \frac{3}{4}\frac{e\,\hbar\,F}{\sqrt{2\,m_r^*(\hbar\omega - E_G)}}, \quad (11.29)$$

which decreases with increasing energy and depends linearly on the electric field. The amplitude of the oscillations has the form

$$4\left(\frac{\hbar\omega - E_G}{E(F)}\right)^{3/2} = 4\,\frac{\sqrt{2\,m_r^*}}{e\,\hbar\,F}\,(\hbar\omega - E_G)^{3/2}, \quad (11.30)$$

i.e., the amplitude decreases as $(\hbar\omega - E_G)^{3/2}$ and increases linearly with the electric field strength. In the limit of very large energies, the absorption coefficient approaches the absorption coefficient for free carriers without any electric field (cf. Eq. 10.20).

For energies below the energy gap, we have to use Eq. 11.11 as the asymptotic limit. The corresponding derivative of the Airy function in Eq. 11.26 reads

$$\lim_{x\to\infty} \mathrm{Ai}'(x) = -\frac{x^{1/4}}{2\sqrt{\pi}}\left[1 + \frac{21\,c_1}{10}\,x^{-3/2}\right]\exp\left(-\frac{2}{3}\,x^{3/2}\right). \quad (11.31)$$

The resulting absorption coefficient below the energy gap is of the form

$$\alpha(\hbar\omega,F) = \alpha_0^{3D}\,\frac{e\,\hbar\,F}{8\sqrt{2\,m_r^*}}\,\frac{\Theta(E_G - \hbar\omega)}{E_G - \hbar\omega}$$
$$\times\exp\left(-\frac{4\sqrt{2\,m_r^*}}{3\,e\,\hbar\,F}\,(E_G - \hbar\omega)^{3/2}\right). \quad (11.32)$$

In this case, the absorption coefficient α exhibits an exponential dependence on both, the energy and the electric field. A physical picture of the finite absorption coefficient below the energy gap is the existence of photon-assisted interband tunneling, when the energy bands are tilted by the electric field. Plotting the absorption coefficient on a semi-logarithmic scale vs $(E_G - \hbar\omega)^{3/2}$ results in a straight line with a slope s inversely proportional to the electric field strength, i.e.,

$$s = \frac{4\sqrt{2\, m_r^*}}{3\, e\, \hbar\, F} \,. \tag{11.33}$$

In Fig. 11.1 the three-dimensional absorption coefficient of free carriers without (dashed line) and with (solid line) an electric field is shown over a large energy range near the energy gap. The parameter $E(F)$ has been set to 0.1 eV corresponding to an electric field of 380 kV cm^{-1} for GaAs. The modification of the absorption coefficient in the vicinity of the fundamental

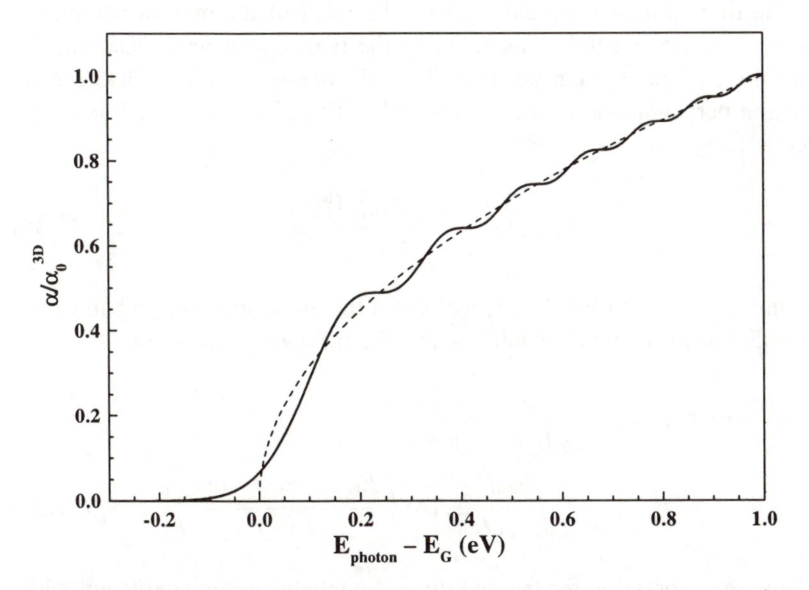

Fig. 11.1. Absorption coefficient of free carriers without (dashed line) and with an applied electric field (solid line) for $E(F) = 0.1$ eV in a three-dimensional semiconductor.

band edge by an electric field is known as the Franz-Keldysh effect, the oscillations above the energy gap are the Franz-Keldysh oscillations.

2. Electroabsorption of Free Carriers in Lower Dimensions

We will now discuss the effect of an electric field on the absorption coefficient of a two- and one-dimensional system. We will first describe the Franz-Keldysh effect in lower dimensions, where the electric field is applied parallel to a direction of the low-dimensional system. The second part is devoted to the effect of an electric field applied perpendicular to the system, the so-called quantum-confined Stark effect.

2.1. Electric field applied parallel to the low-dimensional system

The modification of α by an electric field in the plane of the two-dimensional system can be derived directly from Eq. 11.21. In contrast to the three-dimensional case, where the DOS of the motion perpendicular to the electric field was given by the two-dimensional DOS, for the two-dimensional system we have to use the one-dimensional DOS for the motion perpendicular to the electric field. Therefore, Eq. 11.22 has to be replaced by

$$g(E_{k_\parallel}) = \frac{\sqrt{2\, m_r^*}}{\pi\, \hbar} \frac{\Theta(E_{k_\parallel})}{\sqrt{E_{k_\parallel}}} . \tag{11.34}$$

Using this result in Eq. 11.21, we obtain for the absorption probability in two dimensions with an electrical field the following expression

$$\begin{aligned} \alpha(\hbar\omega) &= \frac{\alpha_{pre}}{z_0\, E(F)} \frac{\sqrt{2\, m_r^*}}{\pi\, \hbar} \\ &\times \int_0^\infty \frac{dE_{k_\parallel}}{\sqrt{E_{k_\parallel}}} \left| \mathrm{Ai}\left(\frac{E_{k_\parallel} + E_G - \hbar\omega}{E(F)} \right) \right|^2 . \end{aligned} \tag{11.35}$$

Using the expression for the two-dimensional absorption coefficient without any electric field in Eq. 10.43 and a partial integration similar to the three-dimensional case in Eq. 11.25, we can write

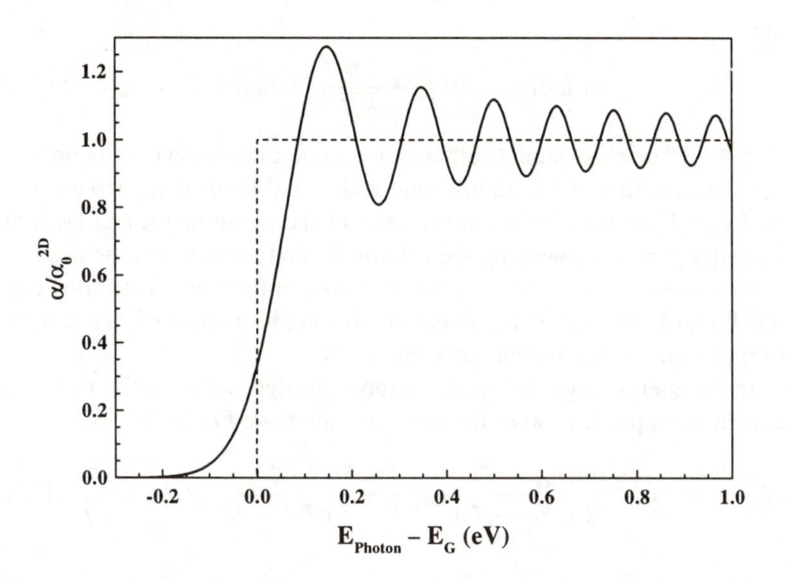

Fig. 11.2. Absorption coefficient of free carriers without (dashed line) and with an applied electric field (solid line) for $E(F) = 0.1$ eV in a two-dimensional semiconductor.

$$\alpha(\hbar\omega) = -8\,\alpha_0^{2D} \int_{x_0}^{\infty} dx\,\sqrt{x - x_0}\,\mathrm{Ai}(x)\,\mathrm{Ai}'(x)\,. \qquad (11.36)$$

The joint density of states without (dashed line) and with (solid line) an electric field is shown in Fig. 11.2 for $E(F) = 0.1$ eV. Below the energy gap, there is again an exponential tail, while the absorption coefficient above the energy gap oscillates around the constant value of the absorption coefficient without any electric field.

For an electric field applied parallel to a one-dimensional system, we can directly use Eq. 11.20 (without the integral over $E_{k_{\|}}$) to obtain the following absorption length

$$\alpha(\hbar\omega) = \frac{2\,\alpha_{pre}}{z_0\,E(F)}\,|\mathrm{Ai}(x_0)|^2\,, \qquad (11.37)$$

where x_0 is given in Eq. 11.24. Using the expression for the one-dimensional absorption coefficient without any electric field from Eq. 10.44, we can write

$$\alpha(\hbar\omega) \;=\; \alpha_0^{1D}\,\frac{2\,\pi}{\sqrt{E(F)}}\,|\mathrm{Ai}(x_0)|^2\,. \tag{11.38}$$

In Fig. 11.3, the absorption coefficient for a one-dimensional semiconductor is shown without (dashed line) and with (solid line) an applied electric field for $E(F) = 0.1$ eV. The divergence of the one-dimensional DOS at the energy gap is removed by the electric field. Furthermore, the absorption coefficient becomes zero for certain values of the applied electric field. In the limit of very high energies or very low fields, it approaches again the absorption coefficient without an electric field.

In the energy range below the absorption edge, we obtain in this case using the asymptotic limit of the Airy function from Eq. 11.11

$$\alpha(\hbar\omega) \;=\; \alpha_0^{1D}\,\frac{\Theta(E_G - \hbar\omega)}{2\sqrt{E_G - \hbar\omega}}\exp\left(-\frac{4\sqrt{2\,m_r^*}}{3\,e\,\hbar\,F}\,(E_G - \hbar\omega)^{3/2}\right) \tag{11.39}$$

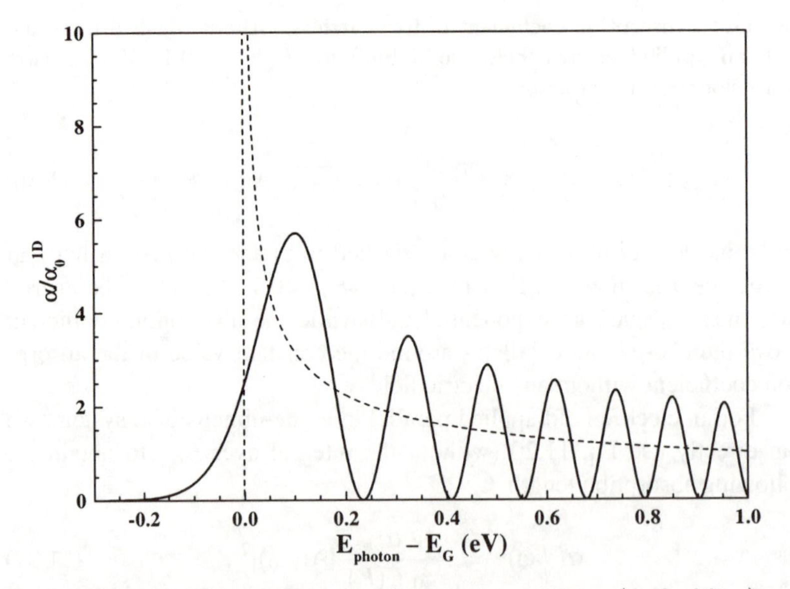

Fig. 11.3. Absorption coefficient of free carriers without (dashed line) and with an applied electric field (solid line) for $E(F) = 0.1$ eV in a one-dimensional semiconductor.

For energies above the absorption edge, we have to use the asymptotic limit from Eq. 11.12 resulting in

$$\alpha(\hbar\omega) = \alpha_0^{1D} \frac{\Theta(\hbar\omega - E_G)}{\sqrt{\hbar\omega - E_G}} \left[1 - \sin\left(\frac{4}{3} \left(\frac{\hbar\omega - E_G}{E(F)} \right)^{3/2} \right) \right] . \quad (11.40)$$

2.2. Electric field applied perpendicular to the low-dimensional system

When the electric field is applied perpendicular to the two-dimensional system, the discrete eigenstates of the 2D-system are changed, while the in-plane motion remains unchanged. In a quantum well with infinitely high barriers, the eigenstates are given by

$$E_z^n = \frac{\pi^2 \hbar^2}{2 m^* d^2} n^2 , \quad (11.41)$$

where d denotes the width of the quantum well and $n = 1, 2, 3, \ldots$. The total energy of the 2D-system is then given by

$$E_n(k_x, k_y) = \frac{\hbar^2 (k_x^2 + k_y^2)}{2 m^*} + \frac{\pi^2 \hbar^2}{2 m^* d^2} n^2 . \quad (11.42)$$

If the origin ($z = 0$) of the quantum well is chosen at its center, the eigenfunctions can be grouped in terms of the parity

$$\Psi_n(z) = \sqrt{\frac{2}{d}} \cos\left(\frac{n \pi z}{d} \right) \text{ for } n \text{ odd and}$$
$$\Psi_n(z) = \sqrt{\frac{2}{d}} \sin\left(\frac{n \pi z}{d} \right) \text{ for } n \text{ even} . \quad (11.43)$$

If an electric field is applied perpendicular to the quantum well, i.e., in z-direction, the energies of the states can be calculated by perturbation theory using the unperturbed wave functions of Eq. 11.43 and the perturbation operator $-e\, F\, z$. The term in first order $-\langle \Psi_n(z) | eFz | \Psi_n(z) \rangle$ vanishes, since the wave functions have a definite parity and z is of odd parity. The change of the ground state energy is therefore determined by second order perturbation theory

$$\begin{aligned} \Delta E_z^1 &= E_z^1(F) - E_z^1(0) \\ &= e^2 F^2 \sum_{m \neq 1} \frac{|\langle \Psi_1(z)|z|\Psi_m(z)\rangle|^2}{E_1 - E_m}. \end{aligned} \quad (11.44)$$

The dominant term originates from the interaction with the first excited state, i.e., $m = 2$. If we evaluate this term, we obtain

$$\begin{aligned} \Delta E_z^1 &= e^2 F^2 \frac{|\langle \Psi_1(z)|z|\Psi_2(z)\rangle|^2}{E_1 - E_2} \\ &= -\frac{1}{2\pi^6} \left(\frac{4}{3}\right)^5 \frac{e^2 m^* F^2 d^4}{\hbar^2}. \end{aligned} \quad (11.45)$$

The prefactor $\left(\frac{4}{3}\right)^5/(2\pi^6)$ in this equation, which is obtained by taking only the first excited state into account, is equal to 2.1916×10^{-3}. If the full series in Eq. 11.44 is considered, the prefactor changes to $\frac{1}{24\pi^2}\left(\frac{15}{\pi^2} - 1\right) = 2.1945\times10^{-3}$. The difference between these two values is about 3×10^{-6}, which is three orders of magnitude smaller than the value of the prefactor in Eq. 11.45, so that only the first excited state is of importance for the determination of the energy shift.

Eq. 11.45 is only valid in the regime $\Delta E_z^1 \ll E_z^1$. In order to determine the energy for higher fields, the Schrödinger equation has to be solved using the Airy functions. In most cases, the field dependence of the energies is determined by a numerical solution of the Schrödinger equation. The effect of the electric field on the absorption spectrum is two-fold. First, the energy gap red-shifts. Combining the energy change of the conduction and valence band states, the energy gap in the electric field becomes

$$E_G(F) = E_G(0) - \left(\frac{4}{3}\right)^5 \frac{e^2 (m_e^* + m_h^*) F^2 d^4}{\pi^6 \hbar^2}. \quad (11.46)$$

The magnitude of the shift depends quadratically on the electric field and linearly on the effective mass. Therefore, for a given field and thickness d, the larger contribution to the red-shift originates from the valence band states due to the heavier mass of the holes. Due to the d^4 thickness dependence, the energy shift is negligible for very narrow quantum wells and becomes much larger for wide quantum wells. The second effect is a change of the oscillator strength. While allowed transitions at zero electric field become partially forbidden, transitions with zero oscillator strength at zero electric field, i.e., forbidden transitions, become allowed for finite electric field with an oscillator strength, which is sometimes even comparable to the one of allowed transitions at zero electric field. This change in the oscillator strength is due to the fact that the electric field removes the definite parity of the wave functions. Piezoelectric semiconductors can exhibits quite large internal electric fields so that this effect may even be observed without applying any external field.

In the general case, the Schrödinger equation including the electric field has to be solved, i.e.,

$$\left(-\frac{\hbar^2}{2\,m^*}\frac{d^2}{dz^2} - e\,F\,z + V(z)\right)\Psi_n(z) \;=\; E_z^n\,\Psi_n(z)\,. \quad (11.47)$$

The wave functions are linear combinations of the Airy functions $\mathrm{Ai}(x)$ and $\mathrm{Bi}(x)$. The term $|\Psi(0)|^2$ in the definition of the absorption coefficient has to be replaced by an average over the quantum well volume, i.e., by

$$\left|\int_{-d/2}^{d/2} dz\,\Psi_e(z)\,\Psi_h(z)\right|^2 , \quad (11.48)$$

where $\Psi_e(z)$ and $\Psi_h(z)$ denote the normalized wave functions of the electron state in the conduction band and hole state in the valence band, respectively. The absorption coefficient for the ground state becomes

$$\begin{aligned}\alpha(\hbar\omega) \;&=\; \alpha_0^{2D}\int_0^{\infty} dE_{k_\parallel}\,\delta(E_{k_\parallel} + E_G - \hbar\omega + E_e^1 + E_h^1)\\ &\times\left|\int_{-d/2}^{d/2} dz\,\Psi_e^1(z)\,\Psi_h^1(z)\right|^2 . \quad (11.49)\end{aligned}$$

In the limit of zero electric field, it approaches Eq. 10.43 for a single subband. When other subbands are included, we have to sum over all the subbands involved. The expression for the wave function has to be evaluated numerically. The above considerations can also be applied to one-dimensional systems with the electric field applied perpendicular to system.

3. Excitonic Effects in the Electroabsorption Coefficient

We will now discuss, how the electric field influences the excitonic absorption coefficient. First, we will summarize the behavior in three dimensions. In the last part, we will give a brief review of the effect of the electric field in lower-dimensional systems.

3.1. Three dimensions

The potential energy of the electron-hole pair includes now the Coulomb attraction term. The Schrödinger equation to be solved is of the form

$$\left(-\frac{\hbar^2}{2\,m_r^*}\,\Delta_{\underline{r}}\,-\,e\,F\,z\,-\,\frac{e^2}{4\pi\varepsilon\varepsilon_0|\underline{r}|}\right)\,\Psi(\underline{r}) \;=\; E\,\Psi(\underline{r})\,. \quad (11.50)$$

There are two approaches. In the first one, we treat the external electric field as a perturbation, i.e., it is valid for small fields. In this limit, the binding energy of the exciton will be changed by the quadratic Stark effect (second order perturbation theory). The linear Stark effect (first order perturbation theory) does not give any correction to the ground state energy. Using the expression for the Hydrogen atom for the quadratic Stark effect, the energy correction for the exciton can be written as

$$\Delta E_1 \;=\; E_1(F)\,-\,E_1(0) \;=\; -\frac{9}{8}\,\frac{e^2\,a^{*2}\,F^2}{R^*} \quad (11.51)$$

so that the exciton absorption line will be red-shifted. With increasing electric field, the exciton absorption lines will be broadened. For excited states, the degeneracy of the states will be removed by the linear Stark effect resulting for example in a splitting of the $n = 2$ line into two energetically separated lines.

For larger fields, the field cannot be treated as a perturbation anymore. In this case, the binding energy of the exciton exhibits a reduction resulting in a blue-shift of the exciton absorption line. At the same time, it is strongly broadened. At even higher fields, when the potential energy drop over one Bohr radius of the exciton corresponds to the exciton binding energy, the exciton will be ionized. This will take place for a field strength of

$$F_I \;=\; \frac{R^*}{e\,a^*}\,. \quad (11.52)$$

For GaAs the exciton binding energy is 4.4 meV and the exciton Bohr radius 12.5 nm so that the ionization field strength corresponds to a value of 3.5 kV/cm. In order to write Eq. 11.52 in terms of the only relevant parameters for excitons in semiconductors, we use the expressions for R^* and a^* in Eq. 9.7 and Eq. 9.8, respectively, to obtain

$$F_I = 2.57 \times 10^6 \left(\frac{m_r^*}{m}\right)^2 \frac{1}{\varepsilon^3}, \tag{11.53}$$

where F_I is measured in kV/cm. The electric field strengths can vary over a large range from below a kV/cm up to several MV/cm for different materials. Due to the ionization of the exciton in the limit of large fields, the electroabsorption coefficient of excitons reduces to the absorption coefficient of free carriers.

3.2. Lower dimensions

In two dimensions, the application of an electric field within the plane of the system results in qualitatively similar results as in three dimensions. However, when the field is applied perpendicular to the system, the spatial confinement will prevent the ionization of excitons up to much larger field strengths. One can therefore observe much larger Stark shifts for excitons in quantum wells. A detailed analysis requires a substantial amount of numerical calculations.

CHAPTER 12
MAGNETOABSORPTION

The application of an external magnetic field breaks the translational symmetry perpendicular to the field direction in contrast to an electric field, which affects the symmetry parallel to the field. Due to the Lorentz force, the free electron moves in a circular orbit perpendicular to the magnetic field. However, the motion parallel to the magnetic field is not influenced and corresponds to that without a magnetic field. As a consequence of the quantization of electronic states in the direction perpendicular to the magnetic field, the spatial dimension of the density of states of a three-dimensional semiconductor in a magnetic field is reduced by two to a one-dimensional DOS. For a two-dimensional semiconductor with a magnetic field perpendicular to the system, i.e., parallel to the confinement direction, the dimensionality of the density of states is reduced to zero, i.e., it corresponds to a system with complete confinement. In the first part of this chapter, we will discuss the effect of a magnetic field on the absorption coefficient of free carriers in three and two dimensions. The second part deals with the influence of the magnetic field on excitons. In the last part, the influence of the magnetic field on the absorption coefficient of quantum wells will be presented.

1. Magnetoabsorption of Free Carriers

As in the previous chapter, we have to consider first the effect of the magnetic field on the energy eigenstates of the system. Neglecting the Coulomb interaction between electrons and holes, the Schrödinger equation for an electron-hole pair in a magnetic field can be written as

$$\frac{1}{2\,m_r^*}\left(\underline{p} + e\,\underline{A}\right)^2 \Psi(\underline{r}) \;=\; E\,\Psi(\underline{r})\,. \qquad (12.1)$$

The magnetic field, which is related to the vector potential $\underline{A}$ by $\underline{B} = \underline{\nabla} \times \underline{A}$, is assumed to be parallel to the z-direction, i.e., $\underline{B} = B\hat{\underline{z}}$, which implies a vector potential $\underline{A} = -By\hat{\underline{x}}$. Using all this information in Eq. 12.1, we can rewrite it in the following form

$$\frac{1}{2\, m_r^*}\left(\underline{p}^2 + e^2\, B^2\, y^2 - 2\, B\, y\, p_x\right) \Psi(\underline{r}) \;=\; E\, \Psi(\underline{r})\,. \qquad (12.2)$$

Since the magnetic field term of the Hamiltonian does not contain z or p_z, we can separate the motion parallel from the motion perpendicular to the magnetic field, i.e., $\Psi(\underline{r}) = \Psi(x,y)\Psi(z)$, where $\Psi(z)$ corresponds to the wave function of a free particle. As a result, Eq. 12.2 reduces to

$$\frac{1}{2\, m_r^*}\left[p_y^2 + e^2\, B^2\left(y - \frac{p_x}{e\, B}\right)^2\right] \Psi(x,y) \;=\; \left(E - \frac{\hbar^2\, k_z^2}{2\, m_r^*}\right) \Psi(x,y) \quad . \qquad (12.3)$$

Since this differential equation does not contain the variable x, the wave function can be written in the form $\Psi(x,y) = \Psi(y)\exp(ik_x x)$. By substituting this last expression into Eq. 12.3, we obtain

$$\left[\frac{1}{2\, m_r^*}\, p_y^2 + \frac{m_r^*}{2}\, \omega_c^2\, (y - y_0)^2\right] \Psi(y) \;=\; \left(E - \frac{\hbar^2\, k_z^2}{2\, m_r^*}\right) \Psi(y) \quad , \qquad (12.4)$$

where $y_0 = \hbar k_x/(eB)$ and $\omega_c = eB/m_r^*$. This equation is the well-known Schrödinger equation for the one-dimensional harmonic oscillator. The corresponding eigenenergies are given by

$$E \;=\; \frac{\hbar^2\, k_z^2}{2\, m_r^*} + \left(n + \frac{1}{2}\right) \hbar\omega_c\,, \qquad (12.5)$$

where $n = 0, 1, 2, \ldots$. The energy levels of this system are known as the Landau levels. The wave functions are the Hermite polynomials of the harmonic oscillator with the origin y_0, i.e.,

$$\Psi_n(y) = C_n H_n\left(\sqrt{\frac{m_r^*\,\omega_c}{\hbar}}\,(y - y_0)\right) \times\exp\left(-\frac{m_r^*\,\omega_c}{2\,\hbar}\,(y - y_0)^2\right), \tag{12.6}$$

where H_n denotes the Hermite polynomial of degree n and C_n is a normalization constant given by

$$C_n = 2^{-n/2}\,(n!)^{-1/2}\left(\frac{m_r^*\,\omega_c}{\pi\,\hbar}\right)^{1/4}. \tag{12.7}$$

Using the electron effective mass of GaAs in Eq. 12.5, we obtain a magnetic energy $\hbar\omega_c$ at 10 T of 18.4 meV. The heavy hole in the valence band has due to the larger effective mass a much smaller energy of 2.32 meV.

Every Landau level in Eq. 12.5 exhibits a degeneracy factor, which can be determined through the quantity Δk_x, since the energy does not depend on k_x. Using simple arguments, it can be shown that $\Delta k_x = \Delta p/\hbar = m_r^*\Delta v/\hbar$. For a circular motion, $\Delta v = \omega_c L_x/(2\,\pi)$. Therefore, the degeneracy factor g_B is given by

$$g(B) = \frac{\Delta k_x}{L_x} = \frac{e\,B}{2\,\pi\,\hbar} = \frac{m_r^*}{2\,\pi\,\hbar^2}\,\hbar\,\omega_c\,. \tag{12.8}$$

The resulting absorption coefficient of free carriers in three dimensions in a magnetic field is of the form

$$\alpha(\hbar\omega,B) = \alpha_{pre}\sum_{n=0}^{\infty}\int_0^{\infty} dE_z\,g(E_z)\,g(B) \times\delta\left(E_z + (n+\frac{1}{2})\,\hbar\omega_c + E_G - \hbar\omega\right), \tag{12.9}$$

where $g(E_z)$ denotes the one-dimensional density of states as described in Chapter 5 and α_{pre} was defined in Eq. 11.19. Substituting this expression into Eq. 12.9 using the definition of the prefactor from Eq. 10.20 leads to the final expression for the absorption coefficient in a magnetic field

$$\alpha(\hbar\omega,B) = \alpha_0^{3D}\,\frac{\hbar\omega_c}{2}\sum_{n=0}^{\infty}\frac{\Theta(\hbar\omega - E_G - (n+\frac{1}{2})\,\hbar\omega_c)}{\sqrt{\hbar\omega - E_G - (n+\frac{1}{2})\,\hbar\omega_c}}\,. \tag{12.10}$$

The resulting absorption coefficient for $\hbar\omega_c$=20 meV is displayed as a solid line in Fig. 12.1. The 3D-like DOS of the absorption coefficient without a magnetic field indicated by the dashed line is changed into a sum of 1D-like DOS. Each Landau level contributes to sum with one 1D-like DOS at the respective energy of $E_G + (n+1/2)\,\hbar\omega_c$.

In two dimensions, the application of a magnetic field in the direction perpendicular to the layer system reduces the dimensionality of the density of states from two to zero. The corresponding energies are given by

$$E_m^n \;=\; E_m^{co} + \left(n + \frac{1}{2}\right)\hbar\omega_c\,, \tag{12.11}$$

where E_m^{co} denotes the confinement energy of the mth subband in the two-dimensional system. The resulting density of states is zero-dimensional, since the energy spectrum becomes completely discrete. The resulting ab-

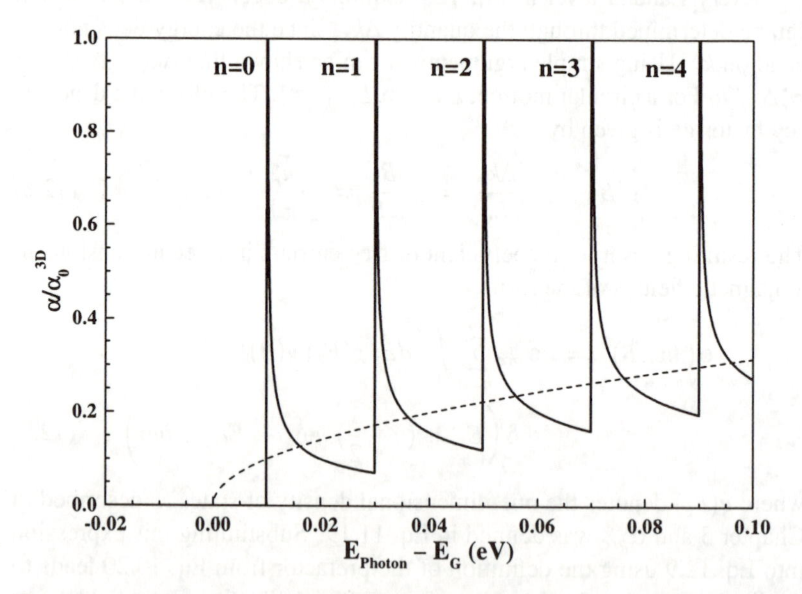

Fig. 12.1. Absorption coefficient of free carriers without (dashed line) and with an applied magnetic field (solid line) for $\hbar\omega_c = 20$ meV in a three-dimensional semiconductor. The respective Landau levels are labeled with index n.

sorption probability is a sum over discrete states. For the lowest subband ($m = 1$), one obtains

$$\alpha(\hbar\omega, B) = \alpha_{pre} \sum_{n=0}^{\infty} g_B \times \delta\left(E_1^{co} + (n + \frac{1}{2})\,\hbar\omega_c + E_G - \hbar\omega\right). \quad (12.12)$$

Rewriting this equation in terms of the two-dimensional absorption coefficient without any field in Eq. 10.43 yields

$$\alpha(\hbar\omega, B) = \alpha_0^{2D} \frac{\hbar\omega_c}{2} \sum_{n=0}^{\infty} \delta\left(\hbar\omega - E_G - E_1^{co} - (n + \frac{1}{2})\,\hbar\omega_c\right). \quad (12.13)$$

The corresponding absorption coefficient for $\hbar\omega_c$=20 meV is shown by the solid line in Fig. 12.2. It was calculated using a line shape function

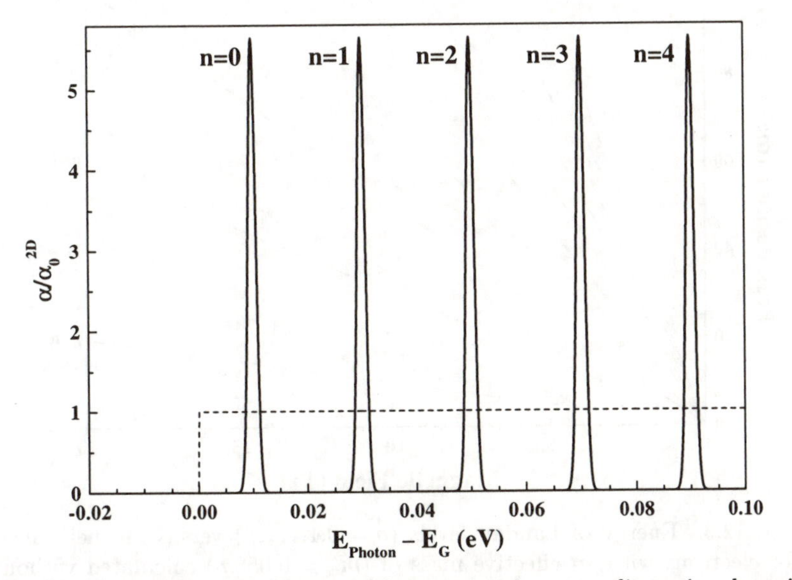

Fig. 12.2. Absorption coefficient of free carriers in a two-dimensional semiconductor without (dashed line) and with a magnetic field (solid line) applied perpendicular to the system for $\hbar\omega_c = 20$ meV. The respective Landau levels are labeled with index n.

$\delta(x) = 1/[\sqrt{\pi}\,\Gamma \exp(-x^2/\Gamma^2)]$ with a line width parameter Γ of $\hbar\omega_c/10$. The confinement energy has been set to zero.

In Fig. 12.3 the magnetic field dependence of the transition energies of the different Landau levels is shown for electrons ($m^* = 0.058m$). The effect of non-parabolicities on the effective mass has been neglected. The plot in Fig. 12.3 is valid for three- as well as two-dimensional semiconductors.

The selection rule for interband transitions in a magnetic field allows only transitions between Landau levels of the same index, i.e., if n denotes the Landau level of the electron and n' of the hole, $n = n'$. If the spin is taken into account, the following additional selection rules apply. If the electric field F_{Photon} of the photon is parallel to the magnetic field, only transitions with $\Delta M_J = 0$ are allowed, while for $F_{Photon} \perp B$ the additional selection rule becomes $\Delta M_J = \pm 1$.

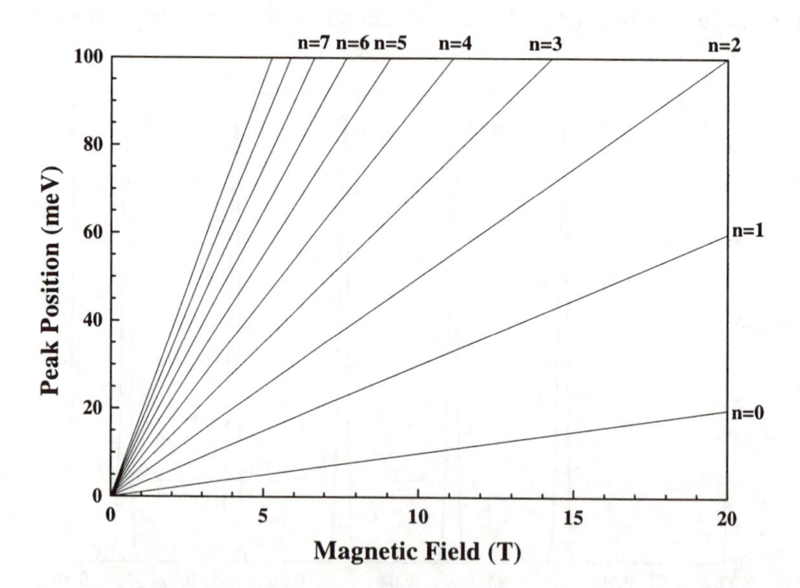

Fig. 12.3. Energy of Landau levels ($n = 0,1,\ldots,9$) versus magnetic field for electrons with an effective mass of ($m^* = 0.058m$) calculated without taking into account any non-parabolicities of the band structure.

2. Three-Dimensional Excitons in a Magnetic Field

The Schrödinger equation for excitons in a magnetic field cannot be solved analytically. As in the case of the electric field, one has to find approximate analytical solutions for the different energy scales. For small magnetic fields, the magnetic energy $\hbar\omega_c$ is much smaller than the exciton binding energy R^*, and the magnetic field can be treated as a perturbation. In the opposite limit, when the magnetic energy becomes much larger than the exciton binding energy, the formation of Landau levels will dominate the spectrum, and excitonic effects are a perturbation. The parameter to be considered is

$$\gamma = \frac{E_{mag}}{R^*} = \frac{\hbar\omega_c}{2R^*} = 4.25 \times 10^{-6} \frac{\varepsilon^2}{m_r^{*2}} B, \qquad (12.14)$$

where B is measured in Tesla and E_{mag} denotes the ground state energy of the lowest Landau level, i.e., $\hbar\omega_c/2$. As long as $\gamma \ll 1$, we can use perturbation theory to determine the effect of the magnetic field on the optical absorption. The perturbation operator for electrons is given by

$$H_B = \frac{1}{2\,m_e^*}\left(2e\,\underline{A}\cdot\underline{p} + e^2\,\underline{A}^2\right) + g_e^*\,\mu_{B,e}^*\,\underline{B}\cdot\underline{S}, \qquad (12.15)$$

where $\underline{S}$ denotes the spin, g_e^* the effective gyromagnetic factor, and $\mu_{B,e}^*$ the effective Bohr magneton of the electron, i.e.,

$$\mu_{B,e}^* = \frac{e\,\hbar}{2\,m_e^*}. \qquad (12.16)$$

The gyromagnetic factor of the free electron is 2. Using a slightly different definition of the magnetic field through the vector potential ($\underline{A} = B(-y,x,0)/2 = \underline{B}\times\underline{r}/2$), one obtains

$$H_B = \frac{e}{2m_e^*}\,\underline{B}\cdot[\underline{r}\times\underline{p} + 2\,\hbar\underline{S}] + \frac{e^2}{8m_e^*}\,[\underline{B}\times\underline{r}]^2. \qquad (12.17)$$

The vector product $\underline{r}\times\underline{p}$ is the angular momentum operator, which can be written as $\hbar\underline{L}$. The second term corresponds to $B^2(x^2+y^2)$. We finally arrive at the expression

$$H_B = \mu_{B,e}^*\,\underline{B}\cdot(\underline{L} + 2\underline{S}) + \frac{e^2\,B^2}{8m_e^*}\,(x^2 + y^2). \qquad (12.18)$$

The first term corresponds to the Zeeman splitting of the states with total angular momentum of $\underline{J} = \underline{L} + 2\underline{S}$. The Zeeman energy splitting is usually written as

$$\Delta E_{Zeeman} = g_e^* \mu_{B,e}^* M_{\underline{J}} B \,. \tag{12.19}$$

If we only consider electrons, $\underline{L} = 0$, the states are only split by the spin. Taking the expectation value of $x^2 + y^2$ with the excitonic eigenfunctions, the second term in Eq. 12.18 describes the diamagnetic shift of the energy levels, i.e.,

$$\Delta E_{Dia} = \frac{e^2 \langle x^2 + y^2 \rangle}{8\, m_e^*} B^2 \,. \tag{12.20}$$

For excitons, we have to combine the contributions from electrons and holes. For the Zeeman splitting, we have to take into account in addition to the spin the angular momentum of the exciton, which is not necessarily zero. However, for the ground state level with $\underline{J} = 0$, the energy is not affected by the magnetic field through the Zeeman effect. Only excited states with $\underline{J} \neq 0$ will be split by the magnetic field.

$$\Delta E_{Zeeman} = \mu_B\, m \left(\frac{g_e^*}{m_e^*} + \frac{g_h^*}{m_h^*} \right) M_J\, B \,, \tag{12.21}$$

where μ_B denotes the Bohr magneton of the free electron. The diamagnetic term can be evaluated for s-states using the relation $\langle x^2 + y^2 \rangle = 2\langle r^2 \rangle / 3$ with the result

$$\Delta E_{Dia}^n = \frac{e^2\, a_n^{*2}}{12\, m_r^*} B^2 = \frac{e^2\, \hbar^2}{24\, m_r^{*2}\, R^*} n^2\, B^2 = \frac{E_{mag}^2}{6\, R^*} n^2 \,. \tag{12.22}$$

The shift of the exciton ground state is therefore

$$\Delta E_{Dia}^1 = \frac{E_{mag}^2}{6\, R^*} \,. \tag{12.23}$$

The diamagnetic shift for the exciton ground state is typically smaller than one meV, since the approximation $\gamma \ll 1$ holds only as long as the magnetic energy is considerably smaller than the exciton binding energy. Since $\hbar\omega_c/2 = 2.1$ meV for GaAs at 2 T, the corresponding shift is about

0.16 meV. For a two-dimensional system, which exhibits a considerably larger exciton binding energy, the diamagnetic shift can be observed up to higher magnetic fields.

3. Two-Dimensional Excitons in a Magnetic Field

We will confine the discussion to a magnetic field applied perpendicular to the quantum well plane. The absorption spectrum of free carriers in a perpendicular magnetic field for a two-dimensional system has already been shown in Fig. 12.2. The step-like absorption spectrum without a magnetic field splits up into a series of sharp lines at the energy position of the Landau levels. The energy gap is blue-shifted by $\hbar\omega_c/2$. A calculation of the absorption spectrum for arbitrary magnetic fields including excitonic effects is beyond the introductory level of this book. However, from the discussion in the previous section, we can conclude that, as long as $\gamma \ll 1$, the excitonic effects dominate over the magnetic confinement. Therefore, the absorption spectrum will be that of the corresponding exciton system as shown in Fig. 12.4 for $\lambda = \gamma = 0.25$. With increasing magnetic field, the Landau levels clearly appear at $\lambda = 1$. The domination of the free carrier Landau levels can clearly be seen for $\lambda = 8$, although the asymptotic limit is not reached at this value, since the oscillator strength of the first Landau level is still larger than the one of the second Landau level. Note that for this value of λ the ratio between the magnetic energy $\hbar\omega_c/2$ is equal to the two-dimensional exciton binding energy $4R^*$. The total blue-shift from purely excitonic absorption to the domination of the magnetic confinement is given by $-4R^* + E_G + \hbar\omega_c/2$. For a value of $\lambda = 8$, the exciton binding energy has just been compensated by the magnetic energy so that the absorption edge occurs at the energy gap of the free carriers.

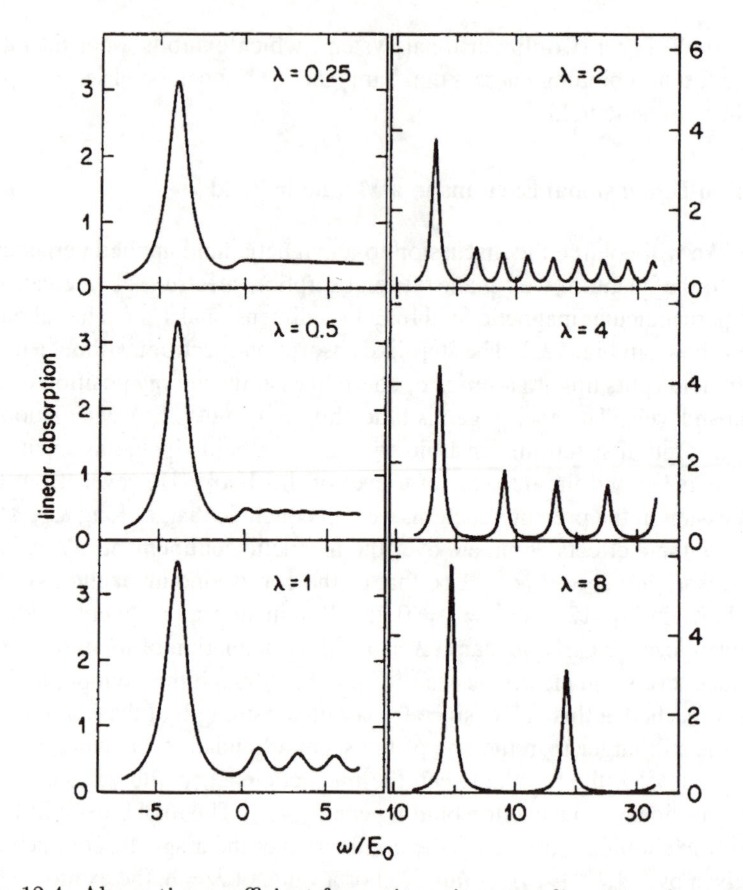

Fig. 12.4. Absorption coefficient for excitons in a two-dimensional semiconductor in the presence of a magnetic field for several values of the parameter $\lambda = \gamma$ taken from *C. Stafford, S. Schmitt-Rink, and W. Schaefer, Phys. Rev. B41, 10 000 (1990).* The energy scale is given relatively to the energy gap for the free carrier absorption. The parameter E_0 corresponds to the 3D-exciton binding energy R^*.

CHAPTER 13
REFERENCES

Condensed Matter Physics

Ashcroft, N.W., N.D. Mermin: *Solid State Physics* (Holt, Rinehart, and Winston, Philadelphia, 1976).

Harrison, W.A.: *Electronic Structure and the Properties of Solids: The Physics of the Chemical Bond* (Dover, New York, 1989).

Ibach, H., H. Lüth: *Solid-State Physics*, 2nd edition (Springer, Berlin, 1995).

Kittel, C.: *Introduction to Solid State Physics*, 7th edition (Wiley, New York, 1995).

Semiconductor Physics

Askerov, B.M.: *Electron Transport Phenomena in Semiconductors* (World Scientific, Singapore, 1994).

Haug, H., S.W. Koch: *Quantum Theory of the Optical and Electronic Properties of Semiconductors*, 3rd edition (World Scientific, Singapore, 1994).

Phillips, J.C.: *Bonds and Bands in Semiconductors* (Academic Press, New York, 1973).

Ridley, B.K.: *Quantum Processes in Semiconductors*, 3rd edition (Clarendon Press, Oxford, 1993).

Sapoval, B., C. Hermann: *Physics of Semiconductors* (Springer, New York, 1995).

Seeger, K.: *Semiconductor Physics*, 6th edition (Springer, Berlin, 1997).

Yu, P.Y., M. Cardona: *Fundamentals of Semiconductor Physics*, (Springer, Berlin, 1996).

Semiconductor Optics

Chow, W.W., S.W. Koch, M. Sargent III: *Semiconductor-Laser Physics* (Springer, Berlin, 1994).

Klingshirn, C.F.: *Semiconductor Optics* (Springer, Berlin, 1995).

Pankove, J.: *Optical Processes in Semiconductors* (Dover, New York, 1971).

Semiconductor Devices

Shur, M.: *Physics of Semiconductor Devices* (Prentice Hall, Englewood Cliffs, 1990).

Sze, S.M.: *Semiconductor Devices, Physics and Technology* (Wiley, New York, 1985).

Semiconductor Parameters

Long, D.: *Energy Bands in Semiconductors* (Wiley, New York, 1968).

Madelung, O., M. Schulz, H. Weiss (eds.): *Landolt-Börnstein*, Series III, Vol. 17a, Group IV and III-V semiconductors (Springer, Berlin, 1982).

Madelung, O., M. Schulz, H. Weiss (eds.): *Landolt-Börnstein*, Series III, Vol. 17b, II-VI semiconductors (Springer, Berlin, 1982).

Madelung, O., M. Schulz (eds.): *Landolt-Börnstein*, Series III, Vol. 22a, Semiconductors (Springer, Berlin, 1987).

Low-Dimensional Semiconductors

Bastard, G.: *Wave Mechanics Applied to Semiconductor Heterostructures* (Halsted Press, New York, 1988).

Grahn, H.T. (ed.): *Semiconductor Superlattices: Growth and Electronic Properties* (World Scientific, Singapore, 1995).

Ploog, K.H. (ed.): *III-V Quantum System Research* (The Institution of Electrical Engineers, London, 1995).

Weisbuch, C., B. Vinter: *Quantum Semiconductor Structures* (Academic, Boston, 1991).

Zhe, C.F. (ed.): *Semiconductor Interfaces and Microstructures* (World Scientific, Singapore, 1992).

CHAPTER 14
FUNDAMENTAL CONSTANTS AND EQUIVALENT UNITS

Fundamental constants

Quantity	SI system	units
Boltzmann's constant k_B	1.38062×10^{-23}	JK^{-1}
Dielectric constant of the vacuum ε_0	8.85419×10^{-12}	$As(Vm)^{-1}$
Electron charge e	1.60219×10^{-19}	As
Electron rest mass m	9.10953×10^{-31}	kg
Planck's constant $\hbar = h/(2\pi)$	1.05459×10^{-34}	Js
Velocity of light c	2.99793×10^{8}	ms^{-1}

Equivalent units

Quantity	SI system	Equivalent
Energy E	10 meV	116 K
	10 meV	2.418 THz
	10 meV	$80.65\ cm^{-1}$
	1 eV	$1.23986\ \mu m$
Phonon frequency ω	1 THz	4.136 meV
	1 THz	47.99 K
	1 THz	$33.356\ cm^{-1}$
Mobility μ	$1\ T^{-1} = 1\ m^2(Vs)^{-1}$	$10^4\ cm^2(Vs)^{-1}$

List of Tables

List of Figures

Index